Cord
810/812
The Timeless Classic

Josh B. Malks

© 1995 by Josh B. Malks

All rights reserved. No portion of this publication may be reproduced or transmitted in any form or by any means, electronic or mechanical, including photocopy, recording, or any information storage and retrieval system, without permission in writing from the author, except by a reviewer who may quote brief passages in a critical article or review to be printed in a magazine or newspaper, or electronically transmitted on radio or television.

Published by

700 E. State Street • Iola, WI 54990-0001
Telephone: 715/445-2214

Please call or write for our free catalog of automotive publications. Our toll-free number to place an order or obtain a free catalog is 800-258-0929 or please use our regular business telephone 715-445-2214 for editorial comment
and further information.

Library of Congress Catalog Number: 95-77313
ISBN: 0-87341-396-2
Printed in the United States of America

DEDICATION

To my sons, Jeff and Dan
who rode with me on the long journey to knowledge

and to
J.K. Howell
dean of Cord scholars
who patiently answered my dumb questions
until I had learned enough to ask smarter ones

this book is dedicated.

In 1951, Dan Post illuminated the world of the Cord automobile with *The Classic Cord,* the first book ever published on the subject of E.L.Cord's two series of cars, the L-29 and the Models 810 and 812. More than twenty years later, Dan expanded the L-29 section into a new volume called *CORD: Without Tribute to Tradition.* He planned to someday issue a second volume on the models 810 and 812, but died before he could accomplish this.

This book carries the title that would have been given to that second volume, and is so named in respectful tribute to Dan R. Post—author, collector, publisher and historian.

CONTENTS

Acknowledgments 6
Introduction 9
Chapter 1: Prelude 11
Chapter 2: Genesis 20
Chapter 3: Front Wheel Drive, Indiana Style 31
Chapter 4: Body by Buehrig 41
Chapter 5: Powered by Lycoming 52
Chapter 6: Year's End 59
Chapter 7: 1935 63
Chapter 8: California, Here We Come! 76
Chapter 9: Countdown 81
Chapter 10: Showtime 93
Chapter 11: Out the Door 103
Chapter 12: To Market, to Market 117
Chapter 13: The Second Season 133
Chapter 14: Of Legroom and Superchargers 161
Chapter 15: A Little Bit Different 175
Chapter 16: For the Record 189
Chapter 17: Under New Management 201
Chapter 18: Hupp and Graham 210
Chapter 19: A-C-D redux 214
Chapter 20: The Enthusiasts 223
Afterword 229
Bibliography 231

Appendices:
i. Production Statistics 233
ii. Reports of the Travels of E306 #2 235
iii. Cord Advertising 240
iv. Numbers 244
v. Paint and Interiors 266
vi. Weights and Measures 268
vii. Cord Speed Records 270
Index 271

ACKNOWLEDGMENTS

It's strange, but true, that the further we get in time from some historical happenings, the more we learn about them. Letters and documents surface from private collections or from the grandchildren of the writers. Previously silent individuals who were actually on the scene speak. Overlooked details are recognized in familiar photographs. And, the perspective of time permits us to evaluate history more clearly.

There hasn't been a lot of in-depth research on the Cord automobile; its creation, short production life, and long journey into the hearts of auto enthusiasts. Much has been written, but all too often the writings have been based on the same terse facts, repeated over and over.

This saga of the Cord is based on the best information available today. It blends the accumulated knowledge of the scholars listed below with my own research and guesses. What you'll read in this book may, in some cases, be at odds with what you've always known to be true. Where I state something as a fact, I have either documentation from the era to back it up, or eyewitness accounts of those who were there, or both. Where a statement is a guess, I say so, and share with you the information that leads me to that conclusion.

Many of the photographs reproduced in this book are quite old. Some are reprinted from early publications, and are not of the best quality. I felt you'd rather see them dimly than not at all.

I share with many of those listed below a nearly lifelong passion for the fascinating Cord 810 and 812. Each of us remembers the first time he or she saw a Cord. I'll bet you do too.

The Historians

Ronald B. Irwin is the conscience of this book. He's been the Cord 810/812 historian of the Auburn Cord Duesenberg Club since 1961. Ron carefully reviewed my drafts. He never let a questionable conclusion pass unchallenged. He forced me to document every statement, and to add disclaimers where I couldn't be certain.

Ron shared unstintingly of his vast collection of Cordiana. (I got a special kick from that very rare occasion when I was able to show him something that he had not seen before!) Whenever I needed accurate information based on printed material of the era, Ron dipped into his magic files and found the reference. The comprehensive list of known Cord numbers is based on decades of Ron's work. The ads pictured in the appendix are from his collection, as is much of the reprinted material. This book would not be what it is without his aid, and I will always be grateful to him.

Henry C. Blommel has secured for Connersville, Indiana, its place in the history of the automobile. Henry located many of the original negatives of early Cord prototypes and production cars. He interviewed employees of the companies that built cars here and recorded some of their firsthand recollections. For his efforts, he deserves the thanks of townspeople and of automotive historians. Henry and I may sometimes differ on conclusions to be drawn, but I am second to none in my admiration for his labors.

The Artist

Ken Eberts is one of the finest automotive artists in the world. He's a founder and past president of the Automotive Fine Arts Society. His work has graced posters, programs and greeting cards, and hangs in homes, offices and galleries. A Cord enthusiast himself, Ken was able to make previously unseen pieces of the Cord story come alive.

The Scholars (in alphabetical order)

Paul Bryant is a past president of the Auburn Cord Duesenberg Club. He interviewed many of the principals in the Cord story years ago, and shared their memories with me.

Randy Ema owns many of the original factory blueprints and sketches pertaining to the Cord 810, and an extensive collection of factory photographs. He was most generous with his time, knowledge and materials.

Bob Fabris was a founder of the Auburn Cord Duesenberg Club. He redrew faded prints so they could be reproduced in this book. His scholarly perspective on history is appreciated, and I treasure my monthly dinners with him.

Stanley Gilliland maintains a unique collection of rare Cordiana, which he shared with me enthusiastically.

Jeff Godshall is a leading scholar of the Hupp Skylark and Graham Hollywood, cars that he calls "the pseudo-Cords." He made certain that I got this part of the story right.

Karl Ludvigsen shared correspondence from his huge library, and brought to light details about Cord engineering that were previously unknown.

Strother MacMinn, designer and teacher, is highly respected by those who style automobiles. His perspectives on the creation of the Cord 810 have been of great value to me.

Rod MacRae owns the most extensive collection of prints of production parts for the Cord 810 and 812. He opened his files to me, and provided data that rounded out many parts of this story.

Don Mates is a veritable encyclopedia of automobile history and other arcane facts. He's known in the old car world as "Dr. Flywheel." I never come away from a conversation with him without having learned something new.

Glenn Pray is a unique individual, whose knowledge of original Cord history is of that special first-person variety. Glenn's heroic efforts helped preserve the Cords he loves. I appreciate the hours he spent helping me get the facts right.

My thanks also to Dennis Bayer, Bruce Earlin, Bill Fay, Lee Foldenauer, Paul Greenlee, Joe Knapp, Everett Kuhn, Jim Lawrence, Paul Lazarus, Bob Mosier, Henry Portz, Don Randall, and the many others whose contributions are mentioned in the text or photo captions.

The Families

Virginia Kirk Tharpe Cord received me at her home on several occasions, and shared personal memories that helped illuminate this story of her late husband's cars.

Barbara Buehrig Orlando diligently searched out notes made by her father, Gordon Buehrig, and corroborated specific references in his diary.

Marvin Jenkins provided photographs not published before, and the opportunity to copy notes made by his father on the backs of his family's copies of the Cord record run photos.

The Institutions

The Auburn-Cord-Duesenberg Museum's renowned collection, and its archivist Gregg Buttermore, were of great assistance in illuminating the story of the cars, the town and the company.

The Companies

Elmer C. Bartels, **Lester See** and **Don Miller** provided accurate details of the operations of the several companies.

Stanley Lidell told me of his father's work with Dallas Winslow and shared some interesting correspondence.

John McQuown was an employee of the Auburn-Cord-Duesenberg Company during the Dallas Winslow period. He helped me contact former employees, and gave me access to much material of that era.

The Suppliers

August W. Rickenbach, formerly of Lycoming, is gifted with a prodigious memory for the happenings of sixty years earlier, and was kind enough to spend hours and hours with me and my tape recorder.

Fred F. Miller, formerly of Gear Grinding Machine Company and the Dana Corporation helped me understand the Rzeppa constant velocity joint.

Eugene Graham, Al Light and Jim Zerfing, present and past employees of Lycoming, filled in other previously unknown details.

The Racing Folks

Chris Economaki of National Speed Sport News publicized my search for information. Jim Hoggatt of the Indianapolis Speedway introduced me to John Laux of Firestone Racing, who led me to Otto Wolfer. Otto provided eye-witness accounts and photos that added new perspectives to the Cord story.

INTRODUCTION

In New York City, the crowds gathered early. The late fall sky was dreary, threatening rain, and the doors of the Grand Central Palace exhibition hall on Lexington Avenue would not open for hours yet. But the mood was festive, considering that the country was still in the depths of the Great Depression. Besides, many in the waiting lines were not employed, and what better had they to do?

In Los Angeles, the sun shone brightly. Crowds arriving by bus and by the big red cars of the Pacific interurban rail system gathered here too, outside the Exposition Center at Beverly Boulevard near Fairfax Avenue. It was easy to wait in the pleasant California weather, and sidewalk hawkers supplied snacks and souvenirs to those who could afford them.

In both cities, the more affluent arrived by taxicab, in their cars or driven by their chauffeurs. What all had come to see was a great leveler; while only a few could buy, everyone could look, admire and dream.

It was Saturday morning, November 2, 1935. Opening day of the great automobile shows, the exhibitions at which the nation's auto manufacturers unveiled their wares for the year to come. The New York Show was sponsored by the Automobile Manufacturers Association, and was the accepted venue for the introduction of new models. (Not for Ford, though. Henry had never joined the AMA, and Ford-built products were displayed at the Hotel Astor several blocks away.)

America's craze for the automobile was at its peak. Since we live today in an era when new car models invade our living rooms in full color and Dolby surround-sound, it may be difficult to imagine the impact of the annual auto shows of the thirties. New styles and engineering innovations were usually kept secret from the public until the official debut at the shows. If cars were to be sold, they had to be shown. For many manufacturers, especially the smaller independents, this first viewing by the public could make or break sales for the year. And for some manufacturers, one more bad year would be the last.

In earlier years the shows had taken place in January. (Indeed, the 1935 shows had been held only ten months earlier.) The administration of Franklin Delano Roosevelt, then well into his first term as president of an economically-battered United States,

grasped at every straw that could help improve the business climate, or at least the public's perception of it. In March 1935, FDR's economists had suggested that an earlier auto show could have a positive effect on industrial production and on employment. The president made the request, and the AMA complied. Alvan Macauley, president of Packard Motor Cars Corporation, was then president of the AMA. As he put it,

> . . . by advancing the beginning of new model production from the first of the year to the fall, more constant employment, spread over more months, for workers directly or indirectly dependent upon automobile manufacture will result.

The all-important 1936 shows in New York and Los Angeles would therefore take place in early November 1935. Shows in Chicago, Buffalo, Cleveland and other cities were to follow in the next weeks.

The earlier date was an inconvenience to most manufacturers, but they rearranged schedules to comply. The citizens of the small Indiana towns of Auburn and Connersville, headquarters and manufacturing facilities of the Auburn Automobile Company, did not yet know that the change of date was eventually to affect their lives, their fortunes, and the indelible mark they were to make on the world of the automobile. On such pegs hang the threads of automotive history, and of these threads our story is woven.

1
PRELUDE

Great quantities of ink and paper have been lavished on the life and works of Errett Lobban Cord. In the 1930s E.L. Cord was referred to as the "boy wonder of the automobile industry." The impression given was that he had leaped from high school onto the entrepreneurial ladder, with never a slip or a look back. That's not exactly how it happened. We are indebted to the

This ad for the Cord L-29 includes one of the most beautiful automobile photos ever taken. (Dwight Schooling)

Top: From the 1925 catalog: an Auburn sedan, the first car produced under Cord's management. (Randy Ema)

Bottom: Errett Lobban Cord in 1936. (Ron Irwin)

meticulous research of Griffith Borgeson, Randy Ema, Bob Fabris, Ron Irwin and others, who have reconstructed the facts of Cord's early life and later accomplishments. The appendix to this book includes sources for more information about this remarkable individual and the companies, communities and people whose lives he affected. I recount in this prelude, in very abbreviated form, those bits of history that the reader will need to know in order to better understand the story that follows: how the Cord 810 and 812 came to be, and then not to be.

Many of the names of individuals and enterprises who appear in this prelude will reappear as part of the story of the Model 810 and 812.

America's automobile industry had not always been centered in Detroit. During the pioneer era of automobile manufacturing, from the turn of the century to about 1920, no fewer than 179 different makes of cars had been built in Indiana. By 1924, though, only the shells of factories remained throughout the state as memorials to a once-optimistic industry.

In 1924 the Auburn Automobile Company was a 21-year-old car manufacturing business located in the town of Auburn, Indiana. What had started as a family enterprise was purchased in 1919 by a group of Chicago financiers including chewing gum magnate William Wrigley, Jr. Five disastrous years later, the bankers' profits and the value of their investment had fallen dramatically. They needed management assistance, and quickly.

In Chicago in 1924, Errett Lobban Cord was the 28-year-old part owner of a distributorship for the Moon automobile. He had not reached this position easily. A native of Missouri, since his teen years he had thrown himself into enterprise after enterprise with great vigor but little success. He had rebuilt, hopped up and sold Model Ts, and engaged in dirt track racing as driver and mechanic. He started a car wash, and operated a small trucking business. He moved his wife and young son between southern California and

Arizona, and back and forth from Chicago, seeking the opportunity that his entrepreneurial soul craved.

The break came in 1920 when the regional distributor of the Moon automobile in Chicago remembered Cord's demonstrated ability as a salesman during one of his Chicago sojourns. (His friend Harold Ames would later describe E.L. Cord as one of the great salesmen of all time). The distributor reached Cord in Los Angeles and offered him an interest in his new expanded territory. A year later Cord had done so well that he went off on his own, acquiring the Moon distributorship for the states of Iowa, Minnesota and Wisconsin.

Having achieved modest financial success in automobile sales, Cord now looked for ways to invest his small capital in an automobile manufacturing enterprise. When he sold a Moon to one of the financiers who owned Auburn, the connection was made. Cord offered the syndicate a management deal, the precise terms of which are unclear today. Griff Borgeson painstakingly reconstructed this deal in his monumental work *Errett Lobban Cord: His Empire, His Motorcars*. Essentially, the financiers lent Cord the money to buy the Auburn Automobile Company, issuing stock to be redeemed over five years. In July 1924, E.L. Cord left his post as vice-president and general manager of his Moon distributorship and became vice-president and general manager of the Auburn Automobile Company.

Top: 1928 Duesenberg Model J, body by Derham. (Don Howell)

Bottom: The Connersville industries that became part of Cord's auto manufacturing complex. (Henry Blommel)

Top: A Miller front wheel drive race car. (Auburn-Cord-Duesenberg Museum)

Bottom: Auburn's administration building, under construction in 1930. (Auburn-Cord-Duesenberg Museum)

Cord worked furiously and non-stop to revive the moribund company. He reestablished relationships with an apathetic dealer network. He sold many of the cars clogging Auburn's lots through wholesale transactions with large dealers. And he began to advertise a new line of straight-eight Auburns, with engines by Lycoming Manufacturing Company of Williamsport, Pennsylvania. Auburn was still in the red by the end of the year, but only by a little.

At the January 1925 auto shows, Auburn offered a lower, longer family sedan, with a choice of six cylinder or straight eight engines. Dealers responded with enthusiasm. Leaving the engineering and body design to his Auburn staff, Cord spent his days making the rounds of suppliers seeking new parts and easy credit. He struck a deal with Central Manufacturing Company of Connersville, Indiana, to build 100 Auburns of the new design. He visited dealers and sold them on linking their future to

Auburn's. He never stopped selling; by the end of 1925, Auburn had sold over $7,000,000 worth of cars. Sales continued to soar: $10,000.000 in 1926, $14,000,000 in 1927, $16,000,000 in 1928. Auburn's 1929 income was an astonishing $37,000,000 and E.L. redeemed the last of the company's stock from the Chicago financiers. Cord was the new president of the Auburn Automobile Company. He and Auburn now had the cash with which to go shopping.

In 1926 Cord bought all the assets of Duesenberg Motors of Indianapolis in bankruptcy court. With them he gained the services of that unique engineer and mechanical genius, Fred Duesenberg. (Younger brother Augie continued to operate the Duesenberg brothers' racing operations as a separate business. After Fred's death his talents were used for special corporation projects from time to time.) In 1928, Fred Duesenberg's translation of E.L. Cord's concept was announced to the world as the magnificent Duesenberg Model J.

Through 1927 and 1928 Cord acquired vacant industrial buildings in Connersville that had been used to produce cars and components during Indiana's auto-manufacturing heyday. Largest of the buildings in the complex was Central Manufacturing Company, earlier supplier of some Auburn bodies.

In 1928 Cord bought control of Auburn's engine supplier, Lycoming, and made it a subsidiary of Auburn. With it came Spencer Heater, a subsidiary of Lycoming. In 1929 Cord took over Stinson Aircraft and Columbia Axle.

Also in 1929, young Mr. Cord created the corporation that bore his name. Its officers and executives included names that appear over and over in the histories of Cord's enterprises: Manning, Pruitt, Ames, Beal, Willis. Primarily a holding company, at its peak the Cord Corporation owned or held a controlling interest in over 156 companies, most of them in transportation-related industries. Later that year, Cord broke ground for Auburn's splendid new showroom-administration building on South Wayne Avenue in Auburn.

Top: The L-29's engine, drive train and suspension. (Auburn-Cord-Duesenberg Museum)

Bottom: Cornelius Van Ranst. (Auburn-Cord-Duesenberg Museum)

Few who love the automobile would be able to resist the opportunity to create a new make of car crowned with their own name. In this respect E.L. Cord was as human as the rest of us. But because he was E.L. Cord, the car would have to be very special. In 1927, Harry Miller's front wheel drive race cars were the most famous machines on the tracks. Auto manufacturers looked for ways to capitalize on the free publicity. General Motors bought two Millers. Cliff Durant, son of the founder of General Motors had a front drive project going. Packard signed a consultation contract

Errett and Virginia's Cord's mansion in Beverly Hills. (Ron Irwin)

with Miller. Another front drive system was being rushed to production as the Ruxton automobile.

Cord signed a five-year agreement with Miller, offering a salary plus royalties on every production car sold. He added outside consultants to his payroll, to be sure that he won the race to be first in production with an American front drive car. Engineer Cornelius W. Van Ranst was hired to direct the project. Leo Goossen, later to become famous for his work with the Miller/Offenhauser/Meyer-Drake Indianapolis race car engines, worked with Van Ranst on the design. Auburn engineer Harry Weaver was also involved. Famous race car drivers were engaged to test and publicize the new car. To all these talents were added those of Auburn's quietly professional Chief Engineer, Herbert C. Snow.

The new car was to be called the Cord Front Drive. (It's most often referred to as the L-29. That appellation was not used by the factory, except in some internal memos. Enthusiasts hung the label on the car later.) By management order, the new car was to use parts that were available from Auburn bins or that could be purchased from others, wherever possible. Tooling costs were to be kept to a minimum. So while the combination transmission case-bell housing casting was special, the transmission gears were stock Auburn. The engine was a stock Auburn block; Lycoming adapted

the flywheel to the opposite end of the crankshaft.

The basic design of the drive train—quarter elliptic springs, De Dion axle, inboard brakes—all were developed by Van Ranst, with a nod to Miller practice. The prototype was built in the Miller shops in California. The next phase of development took place at the Duesenberg factory in Indianapolis. In August 1928 the project moved to Auburn. Four preproduction cars left Auburn in March 1929 for a shakedown trip to California and back.

Cord won the race for the first front drive car on the American market. The car was a styling sensation. Lower in overall height than any other contemporary car, it changed nearly overnight the popular perception of what an American car could look like. The new Cord was showered with praise during its European debut as well. The graceful lines are draped over a chassis whose unique drivetrain permitted liberties no designer could have taken before. While the basic outline was probably developed by John Oswald, Auburn's body engineer, the final package is believed to be the work of young Alan H. Leamy. Leamy had earlier styled the Model J Duesenberg's factory-supplied hood, radiator and fenders. He was to continue to pen outstanding designs for Auburn for the next several years. Within months of the Cord's introduction, the stock market crashed and the market for automobiles entered a difficult and perplexing decade.

Helen Marie Frisch Cord, who had married Errett in 1914, died in 1930 at the age of 37. Together they had two sons, then aged 15 and 13. Shortly after Helen Marie's death, Cord bought land for a home in Beverly Hills. He married Virginia Kirk Tharpe in 1931. Their mansion was completed in 1932, at the low point of the Great Depression. Years later, Virginia Cord, called Gee-Gee by her friends, named it Cordhaven.

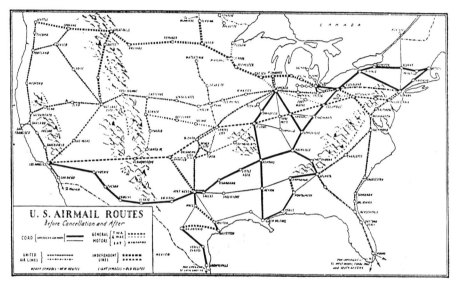

Top: The best selling Auburn ever, the 1931 model. (Auburn-Cord-Duesenberg Museum)

Bottom: Air mail routes, 1933. The legend pits Cord against United and the General Motors group, including TWA.

A Checker Cab built during the years when the company was controlled by the Cord Corporation.

Production of the first Cord Front Drive ended in December 1931, with 5,010 units built over four model years. Auburn's engineers hadn't lost their fondness for front wheel drive, though. Experiments continued.

The Cord Corporation continued to prosper. Cord took over New York Shipbuilding in 1933. (He paid $2 million; a few days later the company was awarded $38 million in government contracts.) Cord acquired control of Checker Cab the same year, and with it the Parmelee System which operated cabs in New York and Chicago. To these he added the Yellow Cab Companies of New York, Chicago and Pittsburgh.

Ranges of over 200 points in a 12-month period were not unusual for Auburn stock during this time. (Eventually they attracted the attention of the new Securities Exchange Commission.) Many years later comedian Groucho Marx would tell interviewers about his grudge against Auburn because their stock gyrations had cost him his investment.

In 1931 Auburn sold 31,000 cars, an astonishing thirteenth in sales of the 32 American car makes. As the depression continued to deepen, though, fewer Americans bought cars. Those who did could afford very expensive machines, or only the cheapest of cars. The battered American middle class, Auburn's customers, had been squeezed from the marketplace.

From its 1931 peak, the Auburn Automobile Company now began its slide toward financial extinction. The Cord Corporation pursued a separate destiny. By 1932 Errett Cord had divested himself and his corporation of most of their Auburn stock. Cord continued to invest, to buy and sell, to make money. But his interest in his eastern and midwestern automotive properties had flagged. By the early 1930s E.L.'s home was in California, and his agile mercantile mind had moved on to other fields, especially the new world of aviation.

Cord had founded Century Airlines in 1930 to provide passenger airliner service and to bid on then-new government air mail contracts. In 1931 *Forbes* magazine reported that Century was carrying nearly one-third of all the business passengers in the United States.

Between 1932 and 1934 Errett Cord appeared twice on the cover of TIME magazine, and on the cover of Forbes. TIME called him "the leading entrepreneur of the Depression." He was described as rarely speaking his mind, "but when he does he uses a language racy and rich with anatomical allusions, forceful expletives."

After a brutal pilot's strike, Cord sold his interests in Century to the Aviation Corporation (Avco) in 1931. Avco called its own airline American Airways. Within one year, Cord and his partners owned a controlling interest. By the next year, following a private and public battle, Cord owned 40% of Avco's stock, and with it American Airways. He had become the dominant figure in an airline industry that had a seemingly limitless future ahead of it. "Over the U.S.," rhapsodized *TIME*, "sweeps the Kingdom of Cord."

In the backwash of the Cord Corporation's successes, the Auburn Automobile Company, which had started it all, struggled on. Sales fell, assets diminished, deficits grew. The corporation's other carmaker, Duesenberg Incorporated, was faring little better. We pick up our story in 1933 Indiana.

Call Him Mr. Cord

Printed references to Errett Lobban Cord use the initials "E.L." as if they were his first name. I asked Virginia Cord once what people called him. She told me that Cord had not cared for the name "Errett" when he was younger, and never used it until he remarried. He liked the way it sounded when she used it, said Virginia, and began to use it regularly. "I called him Errett," she said, "and so did his friends. Close business associates still called him E.L. Everyone else called him Mr. Cord."

For this book I'll use "E.L. Cord" as shorthand for his full name. That's what was printed at the top of his letterhead. He did sign most memos and letters that way, although a photograph I saw in his home was autographed "Errett Cord." So I'll use "Errett" as his first name, but you'll find many uses of "E.L." too. Tradition dies hard.

2
GENESIS

Since its introduction at the automobile salons of 1928, the Duesenberg Model J had been the standard by which luxury cars were judged. All Duesenbergs carried bodies by custom coachbuilders, who vied with each other to produce the most memorable body styles for their wealthy clientele.

In 1933, Duesenberg Model Js were still being sold to the few who could pay for them, although the design was already five years old. The original production goal had been 500 units for just the first year, and this total figure had not yet been reached. Other luxury cars were hedging their bets by offering less expensive models, or by creating a less expensive nameplate that would be associated in the public mind with its luxury stablemate. Packard was working on the Model 120. Cadillac had created the LaSalle. While the Lincoln Zephyr was still three years away, Ford was known to have the car under development.

Duesenberg Model J. (Don Howell)

Harold T. Ames was president of Duesenberg, Inc. As a salesman for Chandler automobiles in Chicago in 1919, he had been a co-worker, friend and roommate of E.L. Cord. When Cord bought Duesenberg in 1926, he appointed Ames sales manager, and later president. As chief executive, Ames ran Duesenberg with vigor and imagination, and was highly respected by his employees.

It was the LaSalle example that Ames had in mind when he conceived the idea for a baby Duesenberg. Cadillac had its corporate cousin Oldsmobile to draw on for a chassis for the LaSalle. Ames had Auburn. A Duesenberg-modified Auburn chassis could be a fine platform for a smaller high-performance vehicle bearing the Duesenberg name. What was needed, as Ames was later quoted, was a "tricky body" for the Auburn chassis.

General Motors dominated the world automobile scene in that era. GM's new unified Art and Color Section was created in 1927 by Harley Earl to make it possible for the body designs of the corporation's cars to be executed on three basic body shells. Among the first designers hired by Earl for his new department was Gordon Miller Buehrig, a young Illinois native with a background in drafting and a burning desire to style automobile bodies. After only a few months working for Earl, Buehrig left GM to work for Stutz Motor Car Company in Indianapolis. In 1929, he applied for the job of chief designer at Duesenberg, and was hired by Harold Ames.

Buehrig drew some of the world's classic automobile styles during these first three years with Duesenberg. In early 1933, he perceived that Duesenberg's prospects were dimming, and sought his old job at General Motors. Earl hired him back.

The memories of Gordon Buehrig and Harold Ames later differed on who called whom to have lunch in the fall of 1933. But they did meet for lunch, at Ames' home in Indianapolis, and they talked about cars. Ames explained his idea for a relatively small, luxury, high-performance car to bolster Duesenberg's flagging fortunes. Buehrig, in turn, described an idea for a sealed engine compartment and radiators mounted outboard in the space between hood and fenders. Ames remembered that Buehrig showed him a sketch of such a design.

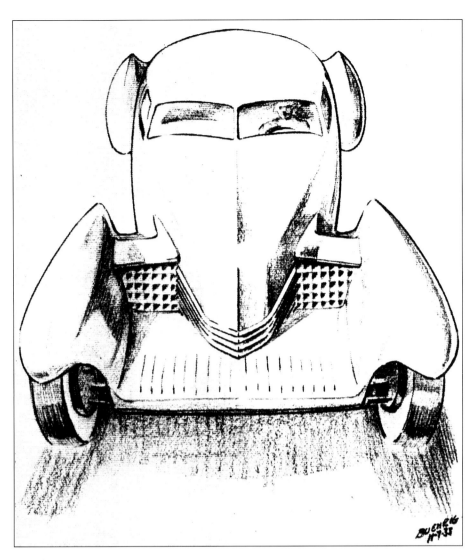

One of Gordon Buehrig's conceptual sketches. He drew this from memory years later. (Auburn-Cord-Duesenberg Museum)

Harold T. Ames. (Auburn-Cord-Duesenberg Museum)

Harold T. Ames

In 1919, after completing his service in World War I, Harold Ames returned to his job selling Chandler automobiles in Chicago. A year later E.L. Cord joined the sales force. He and Ames became friends and roommates.

In 1926, Cord, then president of the Auburn Automobile Company, purchased Duesenberg Motors Corporation, and he offered Ames the position of sales manager. Designing and bringing the Model J to market was the company's main task for the next two years, and Ames was an active participant.

It is to Ames' everlasting credit that he encouraged Gordon Buehrig to return to Duesenberg to develop his new body styling concept. When Ames was appointed executive vice president of Auburn, he brought Buehrig to Auburn.

Ames was an aggressive salesman and executive, and was able to motivate his employees to exceptional performance. Some of Duesenberg's greatest cars were produced during his stewardship, as was the second series of Auburn Speedsters. He did have a tendency to micro-manage. He was not a stylist, but he regularly pushed style proposals on the design staffs. He was not an engineer, but he made suggestions that others struggled to make work. (Ames' idea to adapt the design of the landing light on the Corporation's Stinson planes to the disappearing headlights on the baby Duesenberg was a good one. His convoluted operating mechanism was not.)

Ames' celebrated feud with Roy Faulkner did the Auburn Automobile Company little good. Faulkner supported Buehrig's red clay model as a new venture for Auburn; Ames didn't. But Ames did not use his clout with E.L. Cord to deny life to the project, and he supported it once the decision had been made. For that, enthusiasts should be grateful.

In 1937 Ames left the automobile world but continued his career in industry. He was president of the LaPorte Corporation, which manufactured playground equipment during peacetime, and bomb-bay doors during W.W.II. After the war he headed several other enterprises in the Chicago area. He was a founder of the Mid-America National Bank of Chicago, and sat on the boards of several Chicago corporations.

He died in California in 1983 at the age of 89.

What Ames saw that day was a further development of a concept that Buehrig had executed earlier that year at General Motors' Art and Color Section. It was Harley Earl's practice to encourage competitions among his designers, to develop ideas that might eventually be incorporated into production cars. During a slack period in 1933 such a competition was organized. Attractive prizes were offered to the winners; Earl himself was one of the judges. Four teams of designers were pitted against each other, Buehrig heading one of the teams. Each team created a quarter-scale model of a four-door sedan. As Buehrig later wrote, the judges relegated his team's work to last place. In a private competition engaged in by the designers themselves, Buehrig's design was awarded first place.

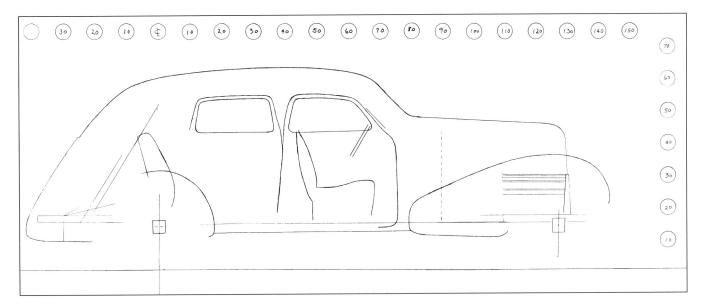

His controversial entry was based on Buehrig's notion that an engine compartment that would stay clean would be a desirable sales concept. Since much of the dirt that covers an engine is drawn in by the radiator fan, the fan and radiator would have to be located outside the engine compartment. Buehrig proposed small radiators to be mounted on each side of the hood.

Ames invited Buehrig to come back to Duesenberg to help turn his sketch into steel. (Buehrig's design work reveals his daring in creating fascinating cars. Barbara Orlando, Buehrig's daughter, says that her father was not a daring man in his personal life. He would certainly not have changed jobs again in that difficult depression year had Ames not held out the strong likelihood of a successful project.) So Gordon Buehrig returned to work for Duesenberg.

A small team worked in secret on the baby Duesenberg project. Buehrig developed the styling in a workroom kept separate from Duesenberg's regular styling studio. Phil Derham, Duesenberg's customer liaison and a designer in his own right, did the body engineering for Buehrig's proposal.

Ames approved Buehrig's pencil sketches in November 1933. The new body had developed more practical "pontoon" fenders and workable proportions. It also had non-functional louvers trimming the front of the hood.

The sketch on page 21 in this book is not what Buehrig showed Harold Ames at their lunch meeting. Ames remembered that the car in the drawing he saw that day had a rounded nose, and no louvers. Nor is the sketch reproduced here one of those that Buehrig presented for Ames' approval in November. Buehrig told me that the originals had long been lost. He created this drawing years later to illustrate his early design concept. Neither the illustration nor the date on it are to be taken literally.

Buehrig developed the design sketches into a practical one-eighth scale drawing, then built a clay model to the same scale. Dimensioned orthographic drawings to guide the body builder were prepared from the model. Phil Derham contracted with coachbuilder A.H. Walker to build the prototype.

Top: A portion of Buehrig's drawing of the baby Duesenberg. (Robert Fabris)

Bottom: Hubert Beal. (Ron Irwin)

Top: The Leamy-designed 1934 Auburn convertible sedan. (Auburn-Cord-Duesenberg Museum)

Bottom: Alan H. Leamy. (Auburn-Cord-Duesenberg Museum)

Albert H. Walker had been a foreman with the Weymann American Body Company of Indianapolis. Many of Weymann's bodies had been built of wood and fabric, in accordance with the patents of its French parent company. By 1931 the market for fabric bodies had all but disappeared. Weymann built a few bodies of more conventional construction, but it was too late. The company closed its doors that year. Walker reorganized it in the same plant as the A.H. Walker Body Company. Walker eventually built six bodies for the Duesenberg Model J, including a very modern coupe styled by chief Duesenberg stylist J. Herbert Newport.

Duesenberg placed its order for the new four-door sedan prototype on January 19, 1934. The order form describes it as an "Auburn Stream-Lined Body." An Auburn eight-cylinder chassis was ordered by Duesenberg on February 5, 1934, and delivered to the Duesenberg factory in Indianapolis. Augie Duesenberg, under a contractual arrangement, worked in an area sealed off from the rest of the plant. His assigned task was to modify the engine and chassis to accept the twin outboard radiators.

Three weeks earlier, Harold Ames had left Indianapolis for Auburn.

L. B. Manning was then vice-president of the Cord Corporation. As trade publication *Automotive Industries* put it, he was "virtually ...in command of the Cord enterprises." Manning and Harold Ames had traveled together by train to and from the 1934 New York Auto

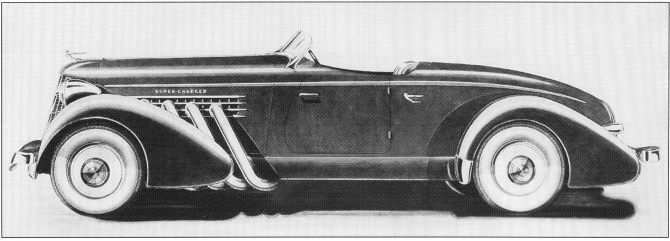

Show. Dealer sales figures generated by the newly-styled 1934 Auburn had not been promising. Ames later said that he had been opposed to the 1934 Auburn body styling from the start, and that Manning now appreciated his perception. More, Manning wanted him to go to Auburn to fix the problem. Ames moved immediately to Auburn, although his formal appointment as Executive Vice-President did not come until later in the year.

At Auburn, Ames found a badly mismanaged company. When Cord had left the presidency of Auburn in 1931, he appointed W. Hubert Beal president. Hube Beal was also president of Auburn's subsidiary, enginemaker Lycoming Manufacturing Company of Williamsport, Pennsylvania. Beal's stewardship at Auburn was marked by optimistic pronouncements that bore little relation to reality. The dealer network was rapidly deteriorating. 1,117 dealers sold Auburn cars in 1932. Only 477 did so in 1934.

Top: 1935 Auburn convertible sedan. The Buehrig changes are evident. (Auburn-Cord-Duesenberg Museum)

Bottom: Lorenzen's illustration of the 1935 Auburn Speedster. (Auburn-Cord-Duesenberg Museum)

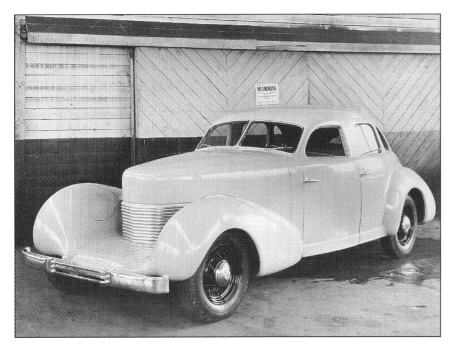

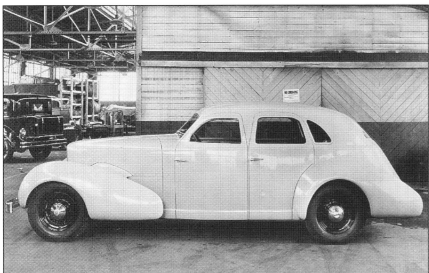

Roy H. Faulkner had left the presidency of Auburn 27 months earlier to first become vice-president of Studebaker Sales Corporation, then vice-president in charge of sales of Pierce-Arrow Motor Car Company. With that proud old firm now on the financial ropes, Faulkner was at liberty. Faulkner was an enormous favorite in the town of Auburn and with the employees of the company, and Manning intended to rehire him as president of Auburn. Ames and Faulkner personally disliked each other. This set the stage for power struggles, with the inevitable detrimental effect on Auburn's employees.

Employee morale was further dampened by the perception of the failure of the 1934 line. Al Leamy, whose genius had been proven by his classic designs for the L-29 Cord and the Duesenberg Model J, had styled the 1934 Auburn. He now found himself caught up in the company's internal conflicts. The 1934 Auburn design was actually an attractive interpretation of the new vogue of streamlining. Time has shown it to be easily up to the standard of its competition. But in that difficult period the company needed someone to blame, and Leamy's design was made responsible for the perceived "disastrous sales" of the 1934 car. Ames had secured his new position at Auburn by advocating change, and change he needed. He decided to immediately redesign the Auburn for 1935, and to introduce the new model in the summer of 1934 instead of waiting for the traditional automobile shows the coming January. (Today that would be called a 1934 1/2 Auburn.) For this he wanted a new designer, so he reached down to Indianapolis for Gordon Buehrig.

Working with Auburn stylist Richard H. Robinson, and within a puny tooling budget of $50,000, Buehrig redesigned the Leamy Auburn's hood, grille and front fenders. A more impressive radiator shell set the tone for a more powerful-appearing hood. The traditional Auburn spear disappeared from the tops of the front fenders. The redesign was completed in a matter of weeks. At the same time, Indianapolis supplier Schwitzer Cummins worked with engine manufacturer Lycoming on the design of a centrifugal supercharger, to be available as an option on eight cylinder

Auburns. Promotional vehicles went on tour in June, and sales to the public began in September.

The early introduction of the 1935 cars left Auburn without a bombshell to introduce at the January 1935 show. To provide such a spectacle, Buehrig and Robinson revamped another Al Leamy design. The last pointy-tailed Auburn Speedster had been offered in the 1933 line. About one hundred unused bodies lay in storage at the plant of the bodybuilder, Union City Body Company. Ames suggested that four of them be used to create attention-getters at the Auburn booths at the shows. The designers adapted the hood and grille of the 1935 Auburn to Leamy's early speedster body shell. Buehrig borrowed the shape of the rear end and fenders from his own magnificent Duesenberg speedster, whose body had been built by Weymann. Buehrig felt that the Auburn actually wound up with the better proportions. The redesign took only two weeks.

In January, the speedsters would be a sensation at the shows. The 1935 Auburns would not. Total sales would reach only 5,063, nearly 500 fewer cars than had been sold in 1934.

Al Leamy left Auburn shortly after Buehrig arrived. His luster had been so dulled by the 1934 Auburn episode, that decades went by before his role in the design of the Cord L-29 and the Duesenberg Model J were recognized. He went to work for Fisher Body in Detroit, and died of an infection a year later at the age of 33.

Left and right: The baby Duesenberg as photographed at the Duesenberg factory in the spring of 1934. (Randy Ema)

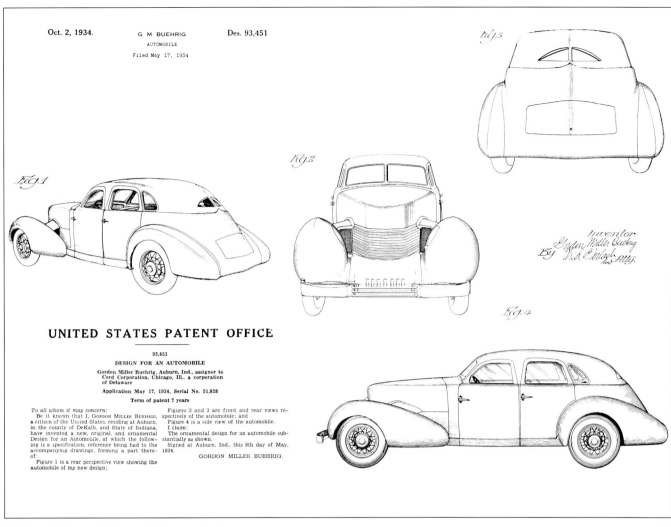

Buehrig's design patent drawings.

> ### Whatchamacallit
>
> Less than three years earlier Auburn had stopped production of the first front wheel drive car in American history to achieve any sales success. In the mind of the car-buying public the tradename Cord was still synonymous with front wheel drive.
>
> The new front drive car is not referred to in writing as a Cord until July 29, 1935. It would be naive, however, to think that Ames did not have that in mind for the car when he shipped the baby Duesenberg off to Auburn. It had to be a foregone conclusion that an Auburn front drive car would be called Cord, and the engineers and stylists must have so referred to it too. "Cord" is a much shorter phrase than "Front wheel drive project E286."
>
> So Cord was probably on everyone's lips, but no one wrote it down. To maintain the historical context, until the car is formally named I'll usually refer to it in this book by its project name.

Buehrig was working on Auburns in Auburn when the new baby Duesenberg body was delivered to the Duesenberg plant in Indianapolis on April 3, 1934. Walker had clearly been instructed to spare no effort to complete the body quickly. The bill for labor alone was $9,035. Considering 1934 wages, Walker's entire crew must have been working on nothing but this car for the ten weeks that the project took. Adding in materials, overhead and profit and taxes, Walker's invoice came to $12,379. The Newport coupe for the Model J chassis, by comparison, had cost just $6,750.

Augie Duesenberg installed the as-yet untested twin outboard radiators on the new Walker body after it had been mounted on the Auburn chassis. Augie later applied for and was granted a patent on the design. The completed prototype was kept behind locked doors, only being brought out to be photographed. Buehrig went back down to Indianapolis to see it.

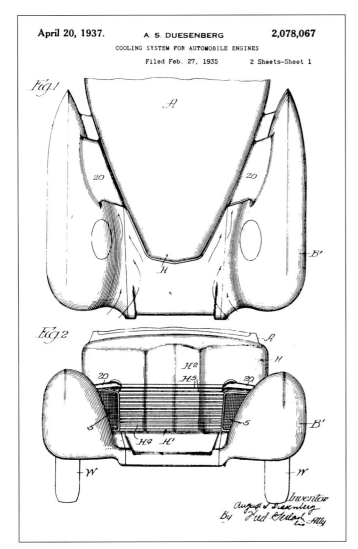
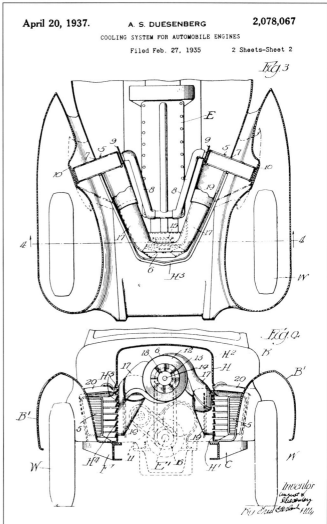

Buehrig had not drawn a bumper for the baby Duesenberg, so Walker didn't make one. Like steering wheels and other hardware, bumpers were supplied to independent auto makers by specialized manufacturers who designed new styles each year to sell to their customers. Salesmen for these manufacturers would visit the styling departments of automakers annually, to show them the new line. One such drummer probably visited Duesenberg in Indianapolis in the spring of 1934. He left a sample front bumper with Augie Duesenberg. It wasn't appropriate for the Model J. But it sure looked right for the baby prototype, so Augie may have mounted it on the car. Buehrig liked it too, and it stayed on. He used photographs of the prototype car to prepare perspective drawings, which were submitted with an application for a design patent on May 17. (Design patent 93,451 was granted on October 2, 1934.)

As new Executive Vice-President of the Auburn Automobile Company, Ames saw himself as having been charged by the top management of the Cord Corporation with the task of saving the company. It apparently took only six weeks for Ames to realize that the challenge of revamping the Auburn line was more important to overall corporate goals than was the creation of a flamboyant small Duesenberg. Ames was also well aware that Auburn's engineers had never really stopped working on front drive designs. The logical

Top: Augie Duesenberg's patent for the outboard radiators.

Bottom: Augie Duesenberg. (Randy Ema)

29

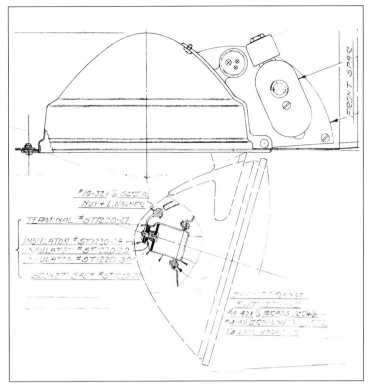 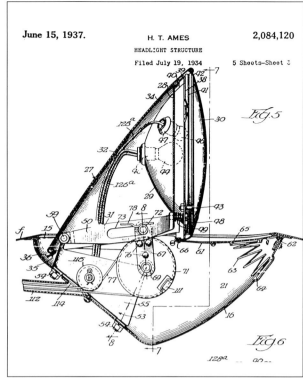

Although they opened in different directions, there's a distinct similarity between the Grimes landing light and the patent drawings for the inboard headlights. The plane used an electric motor and gearset to raise and lower the lights. The car used a patented mechanism of cables and pulleys, and it's this mechanism that proved the undoing of the inboard headlights. US patent number 2,084,120 for the mechanism used on the baby Duesenberg was applied for on July 19, 1934. Harold Ames is named as the inventor. (Paul Greenlee)

move would be to adapt the modernistic design of the Duesenberg prototype to a new front drive car for Auburn. Duesenberg no more, the beautiful vehicle would be turned over to the crew at Auburn.

Gordon Buehrig and Denny Duesenberg, Fred's son, double-dated on Saturday night, June 23. After they took the ladies home they drove to the Duesenberg plant on Washington Street in Indianapolis. They'd been assigned to drive the no-longer-Duesenberg prototype to Auburn. Gordon was driving his customized Ford Model A Victoria, and Denny was to drive the prototype. The distance from Indianapolis to Auburn is about 150 miles. Auburn management wanted the trip to take place in the dark, to preserve the secrecy of the new design. Denny told me that they had to stop every twenty miles for the prototype car to cool down, and he and Gordon would trade off driving the two cars. The sun was coming up when the little caravan finally arrived at the factory in Auburn. They turned the new car over to guards, who had been expecting them hours earlier. Denny remembers that it was too late to go to bed. Gordon had his golf clubs in his Model A, so they went out to the Auburn Country Club and played a round.

3

FRONT WHEEL DRIVE, INDIANA STYLE

Auburn's engineering and design and production staffs kept careful records of their slide rule estimates and actual testing of completed components and cars. Most of this material later went to the Connersville dump. Some of the recorded data did survive, mostly in the notebooks kept by Herb Snow. Not everything was written down, though. Lycoming engineer August Rickenbach told me that when his superior saw him writing up a report after an engine test, he advised him not to waste his time because "nobody was going to read those reports anyway!"

The records that do survive vary widely in quality. Some are carefully annotated graphs of engine performance or other characteristics. A few are handwritten scrawls. Most are signed and dated; some are not. There was, regrettably, a certain looseness about the numerical designations that Auburn gave to its various projects. In some cases projects were given a new number when

The front-drive stub frame, with suspension and steering linkage. Airbrush work is by Paul Lorenzen. (Randy Ema)

Citroen Traction Avant, Model 11CV. (Bob Farrell)

they moved on to their next stage of development. In others, the project number remained unchanged even though the vehicle itself had evolved considerably. Sometimes later documents appear to use the numbers interchangeably! Sorting all this out was a labor of love and exasperation.

Auburn's payroll records could be confusing too. Records refer to the same department by slightly different names in different months. They eventually settled on the names by which I identify them here. The design departments were called "Body & Art Drafting," "Chassis Drafting" and "Design & Research." Creating prototypes was the task of the "Experimental Wood & Metal Shop." Testing was done by the "Experimental Garage."

As projects were assigned by management, each received a number beginning with the letter "E." A "project" could be anything from boat cushions for E.L. Cord's yacht (E257) to the construction of two 12 cylinder 148" wheelbase sedans, with bodies styled by Alex Tremulis (E300). A project might have been a single experimental chassis design like E278, or a series of pre-production Cords like E306. Several projects might be underway simultaneously. Some projects never went beyond the design stage. Some were begun and not finished, while a few made it to prototype stage. Several promising projects might be combined, and given a new E number. Only a very few projects ever entered production. It was Auburn's version of Lockheed Aircraft's later and more famous Skunk Works.

Paul Lorenzen's airbrush rendering of the unit body and bolted-on front stub frame. (National Automotive History Collection, Detroit Public Library)

Project E278 was a "6 cylinder special front wheel drive." Engineer Stanley Lavoie headed this and subsequent front wheel drive chassis development, under the supervision of George H. Kublin. Kublin had been chief engineer of Moon in the 1920s, and had been connected with the manufacture and assembly of the front wheel drive Ruxton automobile. He joined Auburn in 1930.

The first existing document referring to E278 is dated September 7, 1934, with prices compared for Timken and New Departure front wheel bearings. Consideration of a detail of this type would have been preceded by several months of design work, so it's likely that the E278 project began in the spring of 1934. That's right around the time that Harold Ames made the decision that the baby Duesenberg was best developed as a front wheel drive Auburn.

The engineering staff at the Auburn Automobile Company did not design in a vacuum. Herb Snow carefully followed engineering progress by other companies in the United States and Europe. He himself was sought out by motoring journalists for his opinion on future directions of automotive engineering.

Major automobile manufacturers have always purchased competitors' products for testing and analysis. General Motors bought a Cord L-29 in 1929, and a Model 810 in 1936. Citroen of France, probably the world's largest producer of front drive cars in the thirties, bought several Cord 810s.

Auburn bought competitors' machinery too, some of them for front wheel drive studies. Surviving graphs show performance comparisons between Auburn production cars and Willys, Pierce-Arrow and other contemporary makes. Several new cars were acquired in 1934. A DeSoto was driven in a 7-day, 4-state comparison test with Auburn production and experimental vehicles. (One of these boasted 4-wheel independent suspension.) An Oldsmobile 6 cylinder sedan was studied as well. Auburn also purchased and dismantled an Adler, a German front drive car. Surviving drawings show that its suspension design was closely scrutinized. Out of these studies and others came ideas which

Checking Out the Competition

E278 was to use a Continental engine. Did Auburn really consider buying from a competitor to its own subsidiary, Lycoming? Actually, Auburn asked for proposals from other engine manufacturers as well. A letter from Willys-Overland in October 1934 refers to a visit to their plant by Roy Faulkner, and responds to his requests for quotes on four and six cylinder engines. In October 1934 Auburn's graphs show comparative power output of the Continental F-6029 six-cylinder engine and Lycoming's WF series of the same displacement. The Lycoming's output was nearly 10% greater. Continental quoted $60.17 for the assembled engine FOB their plant.

It may have been more practical for Auburn to power experimental prototypes with existing engines, even those of other manufacturers. That would avoid tooling costs for a new engine until decisions had been made. Or perhaps Auburn was just trying to keep Lycoming honest!

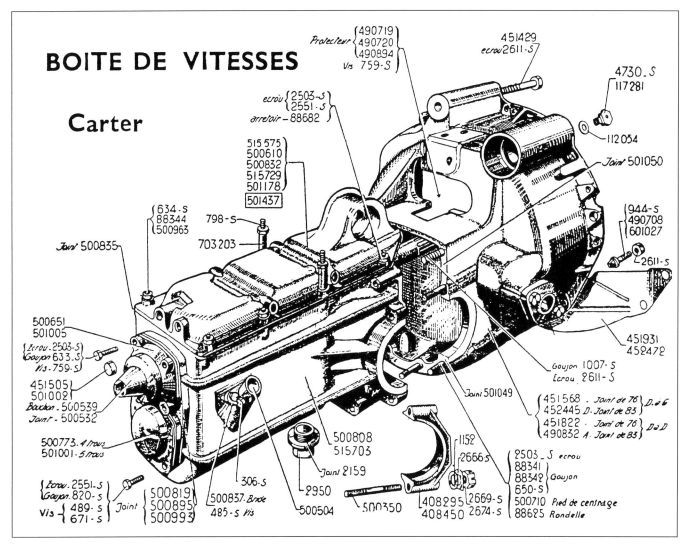

Citroen Traction Avant transmission layout. (Dennis Bayer)

would be incorporated into Auburn's developing front wheel drive chassis.

The first Cord Front Drive, the L-29, had suffered from the management dictum that it was to use as many stock Auburn parts as possible. That resulted in a long drive train—straight eight engine, clutch, transmission, differential—and a weight distribution biased toward the rear. The engineers were not going to make that mistake with the new car. They started with a blank sheet of paper.

The new front drive layout was to be far more compact. The planned V-8 kept engine length down. The drive unit placed the transmission in front of the differential, reducing the length of the package. Most of the weight of the car's mechanical parts would bear on the front wheels. With body, fuel and passengers in place, weight distribution front to rear would be close to an ideal 50-50.

The new car's drive unit—a more accurate term for it than transmission—was probably designed by Auburn engineer Harry Weaver, in collaboration with Detroit Gear & Machine Works. Weaver had been working for Auburn for years, and had been involved in the design of the running gear of the Cord L-29. Dale Cosper, a member of Auburn's styling staff, later remembered him as a thin, reserved gentleman who wore a pince-nez.

Plug and Play

Since nearly all of the car's mechanical parts were mounted on a sub-chassis, Auburn's marketing department had the fanciful notion that this arrangement could give dealers two big advantages over conventional cars. First, if major repairs were required the stub frame could be unbolted, controls unplugged, and a new complete exchange sub-chassis bolted onto the customer's car body. Major overhauls would then inconvenience the car owner only briefly.

Even more bizarre was the idea that dealers could stock three or four sub-chassis' and a variety of car bodies in different models and colors. Chassis would be mated to body when the customer made his final selection. That would reduce the dealer's need to keep a large number of expensive complete cars in stock.

To accommodate the marketing department, the engineers added some quick-release features. The electric shift wiring harness had a plug on the firewall that permited it to be disconnected from the steering column wires. The grille unit was attached to the cowl with pull pins rather than with bolts. It no doubt became painfully obvious that switching frames was by no means a quickie task, and the idea was abandoned. The plug and the pins stayed, though.

The half-baked concept popped up again in the 1948 Tucker. It fared no better this time around.

Auburn engineer Harry A. Weaver, reputed to be the designer of the Cord transmission.

Auburn engineers had a perfect model in front of them for the drive unit of their front drive chassis. They had purchased a new Citroen Model Eleven Traction Avant shortly after it became available to the public in 1934. It was carefully examined, and eventually totally dismantled. A surviving analysis gives the weight of each of the Citroen's components, special attention having been paid to the suspension. The Citroen was much lighter than the new Auburn product would be, and its four cylinder engine of 116.6 cubic inches produced about 46 horsepower. But the drive unit concept was exactly what Auburn needed.

Weaver designed a three-speed transmission and final drive. While the layout differed from Citroen's, the basic ideas are similar. Both designs put the entire gearbox forward of the differential. Both used spiral bevel gears for the final drive. Second and third gears were synchronized in both transmissions.

Rear drive transmissions, then and now, have a "direct drive" in third gear. This means that the driving and driven shaft were locked together in high gear, with the engine power passing through no gears at all. That reduced friction losses and permitted a more compact arrangement of gears.

Like the Citroen, all of the speeds in Auburn's new front drive transmission were indirect. As a result power passed through a set of gears no matter which speed was selected. In high gear the driven shaft and gears in a transmission of this design turn much faster than in a conventional transmission. So special attention needed to be paid to the rigidity of the case in which the shafts were mounted, the bearings in which they turned, and the reduction of friction between the rapidly-spinning gears and the shaft on which they spun.

None of these three-speed transmissions was ever built, and no drawings survive. The engineers experimented with different gear

Another view of the front drive frame. (Auburn-Cord-Duesenberg Museum)

ratios, especially for third gear. The design could not be finalized, however, until a decision was made regarding the displacement of the engine that would power the car.

Universal joint layout was subjected to careful scrutiny, too. A universal joint permits power to be delivered to an output shaft at an angle to the input shaft. Rear drive cars with solid rear axles use one or two universal joints on the driveshaft, since the rear wheels bouncing up and down cause the angle between the transmission and the driveshaft to change constantly. A front wheel drive car uses four universal joints. One is needed on each side of the transmission to accommodate the front wheels moving up and down, and another at each wheel to compensate for steering angles.

A modern front wheel drive car requires that these universal joints be of the constant velocity design; in today's parlance, CV joints. The output shaft on ordinary joints goes through several speed cycles with each rotation. This would create unacceptable oscillation in a front drive car. The design of a CV joint eliminates this cycling.

The two most practical CV joints available in the USA in 1934 had been patented nearly a decade earlier by Carl W. Weiss and Alfred Hans Rzeppa respectively. The Weiss joint was licensed to the Bendix Aviation Company. The Rzeppa joint was manufactured by Gear Grinding Machine Company.

No early decision was made by Auburn on the brand of CV joint to be used. Both designs had to run in a bath of lubricant, so seals became a major design concern. Loss of lubricant would create visible streaks on whitewalls, messy drips on garage floors, and a need for frequent replenishment. None of these would endear the new car to owners. Auburn engineers tried several seal arrangements. (One of the designs they considered for the Rzeppa inner joints included a rubber boot which would turn with the joint. That's exactly the design used on every modern CV joint. Materials of the day were just not up to it.)

The front drive car would have a unit body, essentially welded into a sturdy one-piece box shape from the cowl back. Lancia had pioneered this kind of construction in Europe, and it had been incorporated into prototype cars prepared by Budd and others in the United States. The new Citroen also used a unit body. This design offered the advantages of considerable stiffness and less weight.

Lavoie and his team designed a stub frame to be bolted to the unit body. The frame would carry the engine and drive train, radiator, steering linkage, front suspension and wheels, master cylinder and front brakes, and clutch and brake pedals.

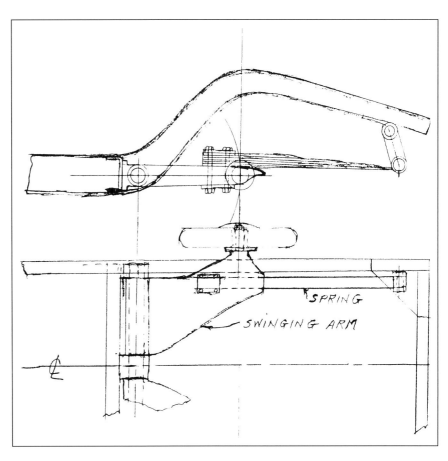

Auburn's experimental independent rear suspension, as drawn by Herb Snow. (Karl Ludvigsen)

One of Herb Snow's pet interests in chassis design was unsprung weight. This is defined as the weight of the parts not carried by the springs. In practice, that usually refers to the wheels and tires, outboard brakes, solid axles, and part of the weight of the suspension arms, steering linkage, swing axles and the springs themselves. Lower unsprung weight means that less shock will be transmitted to the frame and body and that wheels and tires will bounce less, and stay on the road where they belong. Generally, independent suspensions create less unsprung weight than do solid axle designs. Auburn built several rear drive experimental cars with independent front suspensions, and at least one with all four wheels independently sprung.

Kublin and Lavoie worked on the development of the suspension for the front drive chassis. Herb Snow had particular preferences in independent suspension design, and the types of springs to be used. Under his guidance, the Auburn team laid out an independent suspension using a single trailing arm at each corner.

Suspension medium was a transverse spring. Connecting the spring to the suspension arms was a single vertical rod on each side, a design not unlike that used by Corvette in the 1970s. Rubber biscuits, captured in metal covers top and bottom, cushioned the connection at each end. The quarter-elliptic rear springs were attached to the trailing arms in a manner pioneered by the German Adler. Lever action shock absorbers were specified for all four wheels.

Spring rates are measured in cycles per minute. The lower the rate, the softer the spring. Typical spring rates on cars of the 1930s were 150-200 in front and 100-150 in the rear. Rates on the Cord were 72 in

Independent Suspension

Leon Laisne was one of France's more prolific automotive engineers. His designs caught the eye of French auto manufacturer Automobiles Harris in the mid-twenties. A deal was struck. Laisne became Harris' chief engineer. The cars that he designed and they built would carry both names.

The engine of the 1927 Harris-Leon Laisne was purchased from a French supplier. Indeed, for the six years that the firm built automobiles, the engines were always conventional units purchased from others. It was the remarkable chassis design that caused the new car to be widely written up in the automobile magazines of the day. While most production cars were suspended by semi-elliptic leaf springs with solid axles front and rear, the Harris-Leon Laisne offered independent suspension on all four wheels. Short levers on the suspension's swinging arms were linked to coil springs. Automotive journalists were moved to poetry. "The car could be driven over atrocious road surfaces at very high speed without physical discomfort," wrote one. "Exceedingly simple, like so many important inventions," said another. The swing arms were leading at the front, trailing at the rear.

The 1927 Harris-Leon Laisnes were rear wheel driven. By 1932, front wheel drive was used. The springing medium had also changed, from coil springs to rubber. Laisne experimented with different combinations of leading and trailing arms. What didn't change was the basic design of the swing arms. A single forged arm at each wheel pivoted on a tube which was carried in tapered roller bearings.

In 1960 Herb Snow, now chief engineer of Checker Motor Car Corporation, corresponded with Karl Ludvigsen, then Technical Editor of *Car and Driver* magazine. Ludvigsen was preparing an article on cars without driveshafts: front wheel driven, or rear-engined.

In these letters, Snow names the Harris-Leon Laisne as the inspiration for the independent front suspensions with which Auburn experimented. (Experimental rear suspensions in Auburns and the E278 design were based on the Adler car, with leaf springs substituted for the German car's torsion bars.)

Auburn modified the Laisne design as its own needs dictated. The forged swing arm of the much lighter French car would have been extremely expensive to adapt to the heavier Auburn machines. Instead, Snow's men designed hollow swing arms fabricated from stampings. This accomplished the dramatic reductions in unsprung weight that Snow sought, while keeping tooling and manufacturing costs down. Harris-Leon Laisne designs used coil springs or rubber blocks as the suspension medium. Snow preferred leaf springs. He praised them in a letter to *Automobile Topics* in 1935, citing their "natural dampening action" and economical cost. He suggested that they would remain the suspension medium of choice for production cars for the foreseeable future. So Snow's favorite leaf springs and the Harris Leon-Laisne's swing arms set the parameters for the independent suspension of Auburn's new front wheel drive chassis.

Herbert C. Snow (Auburn-Cord-Duesenberg Museum)

Herbert C. Snow

A native of Cleveland, Ohio, Herb Snow graduated from Case Institute of Technology in 1906. He worked for Peerless Motor Company and Willys-Overland; he became chief engineer for Winton in 1914. As a consulting engineer in the early 1920s, Snow worked with Dr. Armand Sperry, inventor of the Sperry gyroscope, on the development of an automatic transmission. He later became chief engineer for Velie Motor Car Company.

In 1927, Herb Snow was appointed vice-president of engineering of the Auburn Automobile Company, where he was heavily involved in the design of the Cord L-29. He was the key figure in the engineering development of the Model 810. Snow was an outstanding engineer who avoided the limelight and freely shared credit with his colleagues.

Snow left Auburn in 1936 to return to consulting engineering work. He moved to Kalamazoo, Michigan, in 1940 to become chief engineer of Checker Motors Corporation. Checker president Morris Markin had a special interest in front wheel drive as a system which could provide more passenger space in a short-wheelbase taxicab. As the American engineer with the most experience with working front wheel drive cars, Snow developed several front drive prototypes for Checker before and after W.W.II. One of these, which did actual service as a taxi, used a four-cylinder transversely-mounted engine, a precursor of the modern front wheel drive car.

Herb Snow still held the position of chief engineer at Checker when he died in 1961 at the age of 76.

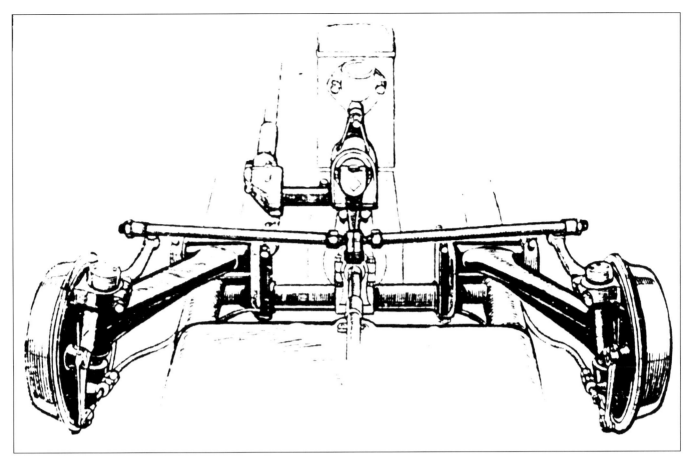

Harris-Leon-Laisne front suspension. While leading arms are used here, Snow used the identical principle for the trailing arms on the front drive car. (Behring Library)

front and 65 in the rear. These rates were much lower than contemporary cars, and predated those used on the softest-riding American cars of the 1950s. While this softer spring permitted a substantial amount of lean on sharp curves, it did contribute to the Cord's outstanding riding qualities, by contemporary standards.

The Bendix duo-servo brake was the latest in brake design in 1934, and the Auburn engineers wanted to use it on their new front drive chassis. Drums 10 inches in diameter and shoes 2 inches wide were specified.

Precision steering was assured by a true center-point linkage design. A forged Y-shaped intermediate arm turned on a double-row ball bearing. Intermediate drag links struck almost exactly the same radii as the suspension arms during suspension movement. Several vendors, including Ross and Gemmer, were considered as suppliers of the steering gear.

Other critical elements in the design of a front drive system are the front hub and front wheel bearings. These bearings carry much greater loads than do the front wheel bearings in a rear drive car, because the drive thrust passes through them as well. In Auburn's new front drive design, compactness was one of the criteria. The engineers wavered back and forth between a pair of tapered roller bearings and a single double-row ball bearing of a new design. If tapered roller bearings were selected, the Timken brand was the only choice since they held the patent on that design. If ball bearings were used, MRC and New Departure each made proposals.

The final decision was based on the compactness of the single ball bearing. Price was not the issue, since the pair of Timken tapered

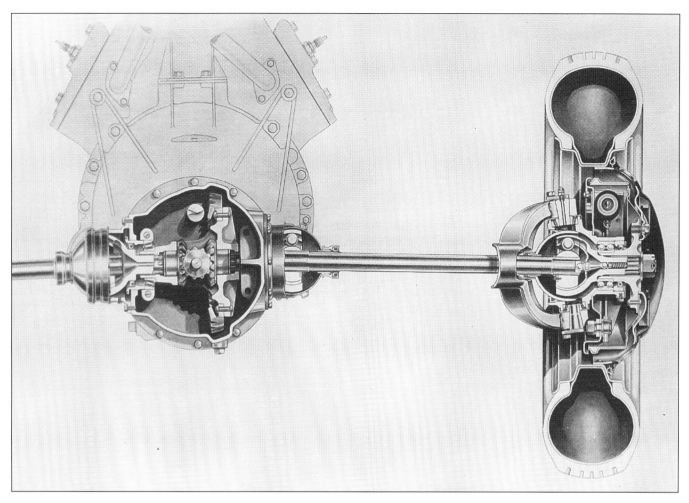

Lorenzen's cross-section of the front drive mechanism. For production the inner universal joints were switched to the Bendix design, and a different lower kingpin was used. (Randy Ema)

roller bearings could be purchased for a few cents less than the ball bearing. The design of the ball bearing put limits on the size of the hub that would fit into it, and set the stage for future problems.

Auburn's extensive experience with front wheel drive showed up in the geometry of the front end design. A drawn line passing through the kingpins intersected the center of the universal joints, and ended within the area or patch of tire contact with the road. This very modern concept was a significant advance over other cars of the period.

The E278's basic mechanical design was even more radical than the Cord that eventually saw production. This car's concept would have been familiar to drivers in the last decade of the twentieth century. E278 was to have been a light 2900 pound front wheel drive car, powered by a 210 cubic inch Continental 6-cylinder engine developing between 71 and 77 bhp, depending on the cylinder head design. Engine and drive train, including a 3-speed transmission, were to be carried in a stub frame, to be bolted to a unit body. Suspension was independent on all four wheels. Wheelbase was to be 120 inches. According to Auburn's graphs, top speed would have been 81 mph.

4

BODY BY BUEHRIG

Gordon Buehrig once said that in the years he worked for Auburn he never attended a committee meeting. Motoring publications have since held up his Model 810 Cord as the last car designed by an individual, rather than by a committee. All this reinforced the notion that the Cord 810 took shape in a single session at Gordon's drafting table, with no input from others. Gordon Buehrig was a modest and gracious man. Many times I heard him lauded within his hearing as the sole designer of the Cord. Many times I heard him reply, "There were several of us who did it." He never failed to use the plural pronoun "we" when recounting how the Cord was designed.

Gordon Miller Buehrig.

Top: Cosper and Richard Robinson. They're using the styling bridge invented at Auburn. (Ron Irwin)

Bottom: Dale Cosper at work in Auburn's styling studio. (Mrs. Dale [Sally] Cosper)

Al Leamy left Auburn shortly after Buehrig arrived at Auburn in 1934. Remaining on the styling staff were Richard H. Robinson and Paul Reuter-Lorenzen.

Robinson was an experienced body draftsman. Buehrig later described him as quiet, unassuming and capable. He had worked with Buehrig on the Auburn redesigns. Reuter-Lorenzen was an outstanding illustrator. His airbrush work was used in some of Auburn's sales material, including much of the information provided in press kits, catalogues and information for salesmen. Dale Cosper remembered that Reuter-Lorenzen drove a Hupmobile roadster with the top down, summer and winter.

Cosper had gone to work for Auburn as a blueprint boy right after his graduation from Tri-State College in Angola, Indiana. He was laid off after about nine months, then rehired soon afterward when he heard that the new head of the Art and Body Drafting department, Gordon Buehrig, was looking for a clay modeler.

Vince Gardner graduated from high school in Duluth, Minnesota, in June 1934, and had come to visit an uncle in Auburn for the summer. Fisher Body Corporation, a division of General Motors, used to sponsor an annual contest for aspiring stylists and modelers. Gardner had won an award for his model of a Napoleonic coach, and showed the coach to Buehrig. He was promptly hired as a model builder.

In July 1934 Buehrig and his team were given their assignment. They were to design a body for a new front wheel drive car, based

Gordon Miller Buehrig

Gordon Buehrig was born on June 18, 1904, the younger of the two children of Fred and Louise Buehrig. After graduating from high school in his hometown of Mason City, Illinois, Buehrig attended Bradley Polytechnic in Peoria. His teachers were unhappy with his notebooks, all of which included sketches of automobile designs.

Buehrig left college in 1924, and worked as an apprentice in the body shop of Gotfredson Body Company in Wayne, Michigan. Over the next several years he held drafting positions with custom bodybuilder Dietrich, Incorporated, and with Packard Motor Car Company. It was while he worked for Packard that a friend got him the interview with Harley Earl that resulted in a job with General Motors' new Art and Color Section. Gordon Buehrig's automobile styling career had begun.

Only a few months later, Buehrig was offered a position as body designer for Stutz Motor Car Company. As he later pointed out, only a very young person would have left the world's largest car manufacturer to work for a small company in desperate financial trouble. Buehrig designed Stutz' Weymann-bodied entries in the 24-hour sports car race at Le Mans, and worked on design details of production cars.

In June 1929 Buehrig was hired by Harold Ames as Chief Designer for Duesenberg, Incorporated. At 25 years of age, he was the head stylist for the most prestigious automobile ever built in the United States. A heady position, and Buehrig rose to it. All Duesenberg Model Js carried bodies built by custom coachbuilders. During Buehrig's three years with Duesenberg, he designed nearly half of those bodies.

In 1933, it was becoming clear that Duesenberg did not have a long-term future. Buehrig wrote to Harley Earl asking about a position. To his own surprise, he was rehired. Months later he left GM again to return to Duesenberg, to execute a new body concept that's referred to today as the baby Duesenberg. Shifted to Auburn, he styled the 1935 and 1936 Auburn line, and the final series of Auburn Speedsters.

Buehrig's masterpiece, the Cord Model 810, was later to make him immortal. He left the Auburn Automobile Company in 1936, before production of the 810 had ended.

In earlier years, the Budd Company of Philadelphia had manufactured bodies for major manufacturers, and sought to break into this business again. Hired as their designer, Buehrig created a new styling studio for Budd. Working with him was Vince Gardner. Buehrig's career at Budd ended abruptly after three years when he complained to the company president about the lack of attention his ideas were receiving.

Buehrig tried some unsuccessful freelance ventures, then moved into aircraft drafting and engineering during World War II. After the war, he joined the Raymond Loewy's Studebaker design group. (Such was the closeness between Buehrig and Gardner, that Vince joined him again in this workplace.) Shades of the old Auburn days, Buehrig again found himself caught in a company crossfire, this time between Raymond Loewy and Virgil Exner, later head stylist for Chrysler. In 1951, Buehrig went to work for Ford Motor Company. He was involved in the design of the Ford Victoria, and in the body engineering for the Continental Mark II. Later he was part of a small group of materials engineers at Ford seeking new ways to use plastics.

Gordon Buehrig retired from his productive automotive career in 1964. Afterward he taught courses at the Art Center College of Design in Los Angeles. He continued to attend the annual reunions of the Auburn Cord Duesenberg Club until his death in 1991 at the age of 86.

Top: Buehrig in his studio in 1935. (Auburn-Cord-Duesenberg Museum)

Bottom: Vince Gardner in the 1930s. (Fran Roxas)

Top: *The red clay model, as it was photographed on July 7, 1934. (Mrs. Dale [Sally] Cosper)*

Bottom: *Paul Reuter-Lorenzen. (Mrs. Dale [Sally] Cosper)*

on the same concept that Buehrig had developed for Ames at Duesenberg. Buehrig emphasized later that the new Cord's body design was not simply a revision of the baby Duesenberg. That vehicle was locked away on a floor below the studio, and was never referred to by the automotive artists. Buehrig began instead with his same germ of an idea that had resulted in the baby Duesenberg project, and developed it in a different direction.

Stanley Lavoie and Auburn's chief body draftsman Ted Allen and his assistant Bart Cotter all wanted to do the new car as a unit body from the cowl back, mating it to a stub frame that would carry the mechanical parts. This package was an ideal one for a body designer to work with. Front wheel drive meant that no provision would have to be made for a driveshaft, so the car could be lower. The unit body permitted frame rails to be built into the perimeter of the body, making step-down design possible. The opportunity to design this new car from scratch, Buehrig later said, was a stylist's dream.

And so the young people of the Body & Art studio treated it. Only Buehrig and Lorenzen were over thirty, and none were married. They worked at their project every day, and most nights. The new concept shaping up in their art studio and on the drafting tables in Auburn's engineering department was generating excitement among young apprentices and seasoned veterans alike. Cooperation was common. When Buehrig asked for changes in the location of the battery and muffler, for example, the engineering department complied.

The Art & Body Drafting studio in early 1935. Left to right: Paul Reuter-Lorenzen, Art Kruger of the Sales Department, Richard Robinson, Gordon Buehrig, Vince Gardner, Dale Cosper. On the rear wall is Lorenzen's full-size airbrush rendering of the new front drive transmission. The model at the extreme right may have been one of Ames' potboilers, or a proposal for a new Auburn. It appears to have been the inspiration for Ab Jenkins' Cord Special. (Ron Irwin)

That doesn't mean that everything was shared. Generally, the engineers told the stylists what they needed to know, and little more. Even in a company the size of Auburn, with all the design work taking place in one building, there was much that the right hand knew that the left one didn't.

Based on the guidelines provided by the engineers, Dick Robinson laid out the parameters of the car's mechanical package on paper, to one quarter of full scale. Project E278 had by now evolved into E286. The latest version would have a 123 inch wheelbase, so that was the dimension that Robinson used.

Now Buehrig had to draw the basic design lines for the body. This orthographic drawing would guide the modelers in the creation of the three-dimensional clay model that would be created next. What he drew was similar to drawing number S10227 shown on page 46. The front edge line of the front door mirrors the line on the photos of the clay model. It is not the same door line as on the prototype Cords that were built later, or on the production 810s.

Buehrig always said that he was a better sculptor than artist. Cosper and Gardner were both modelers. The method of choice in the Auburn styling studio was clearly clay modeling, and the invention that arose from that choice changed the way auto studios modeled their cars for decades to come.

The Body & Art team designed a modeling tool that traveled over the clay model on tracks or guides. Holes drilled on the top and side faces of this bridge accepted dowels whose tips located any

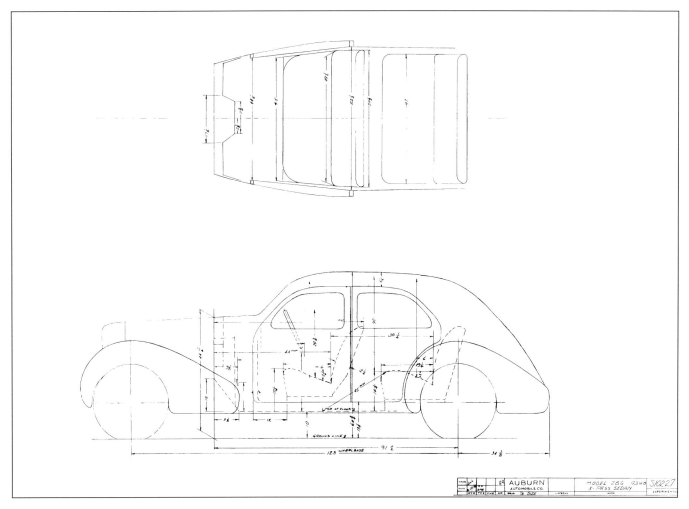

Orthographic drawing of E286. This was probably made just before or just after the red clay model was completed. (Herbert C. Snow collection, Auburn-Cord-Duesenberg Club)

point in space. That meant that after one side of a model was complete, the bridge could be used to make the other side identical. The principle held true whether the model was to scale or full size. So the bridge over the small-scale model could be used to translate the same points to a full-scale clay model that stood under its own larger bridge. Every major styling studio used this bridge until the advent of computer-aided design systems. Buehrig shared credit for this invention with "the genius of Dale Cosper and Vince Gardner."

Gardner built a quarter-scale wood armature with a minimum of one inch left all around for the application of clay. The team laid on the slabs of warm clay. With Buehrig's drawing as the starting point, the work on the model began.

Now the unique and timeless form of what was to become the Model 810 Cord took shape under the talented hands of Gordon Buehrig. He sculpted the body design, based on his own drawings, on one side of the armature. When that side was complete, Cosper and Gardner used the bridge to repeat the shape on the other side. When the job was done, the small group must have stood and gazed in wonder at what they had created.

The next step in the styling of a brand new car design is usually the construction of a full-size clay model. There never was such a model of the Cord-to-be. Buehrig was later to point out that his design team used this to advantage. Hours were spent studying the

> **Design Inspiration**
>
> Later in his life interviewers and Cord fans would besiege Gordon Buehrig, pumping him for information about those heady days when the men of Auburn were making automotive history. Gordon repeatedly pointed out that his work was with the Body & Art section, and he had little knowledge of what was going on in the engineering areas, or in management. During one such session, when the questioners were growing increasingly insistent on knowing exactly what was going on in everyone's mind as they worked, I watched Gordon throw up his hands in exasperation and cry, "I've told you everything I know—I don't know any more!"
>
> Gordon's response was less emotional, but essentially similar, when asked to recollect why he designed a certain line or a certain plane on the car in just that way. He always said that he designed the Cord to look right. When he molded those lines and planes in clay, he did what felt right. The beauty of the Model 810 flowed directly from the inspiration in his fingers. With results like that, who cares why or how?

quarter-scale model from all angles, something that could not easily be done with a full-size car. Despite their tender years, this was a formidable group of connoisseurs of lines and shapes. No doubt suggestions were made for emphasizing a highlight here, improving a line there. Young Vince Gardner was probably especially enthusiastic and involved. As head of the team, Gordon Buehrig listened and made the final decisions. And thus he sealed the Cord's place in the history of automotive fine art.

Dale Cosper remembered that the completed clay model was painted with red lacquer. Later, the clay and paint cracked regularly, requiring quick repairs.

With the shape complete, Gordon and his team could now work on the many details that made up the interior and exterior of the automobile.

From the beginning, Buehrig intended that the car's headlights be retractable. When the early drawings and clay model were created, there was no indication of where the headlights were to be located. The retractable headlights of the baby Duesenberg were located on the inboard surfaces of the fenders. Weymann had made these custom built fenders out of several pieces, carefully welded and leaded at the seams. It was not clear whether it would be practical to fabricate the front fenders of any future production car this way. So the red model suggested two stamped steel sections to be welded down the center, with the weld covered by a chrome molding. With this in mind, the headlights were planned for placement on the inboard surface of the fenders. By the time the first front drive prototypes were built, the fabrication people had found a way to make the front fenders without a visible seam. That made it possible for the headlights to move to the front of the fenders. Still, the first prototypes had inboard headlights.

The bumpers on the new car were not to look anything like those on the red clay model. Rather, they would look like the front bumper of the baby Duesenberg. As the design team worked on details, Buehrig remembered the salesman who had left the sample bumper with Augie Duesenberg. That salesman's company, Buckeye Manufacturing Company of Springfield, Ohio, would later supply the bumpers for the production cars as well.

The interior designs for the new car were based largely on Buehrig's own work at Duesenberg. The use of richly-colored contrasting interior fabrics and paints was new to mass-produced cars, most of which came equipped with mouse-colored mohair upholstery. Buehrig had used this concept in his Duesenberg

Top and bottom: Other views of the red clay model. (Mrs. Dale [Sally] Cosper)

Harold Ames, Stylist

Dale Cosper told Cord scholar Don Mates of Harold Ames' attempts to get Auburn's design team to develop some of his own styling ideas, in preference to the red clay model. As executive vice-president, Ames had the authority to order the stylists to execute his designs. Cosper said that they referred to them as "pot-boilers." Each time they had to do one, they discreetly skewed the proportions so the result looked ungainly. Ames could never figure out why his wonderful ideas all translated so poorly into clay!

When Ames brought Chicago ad agency head P.P. Willis to Auburn to show him the latest designs, Buehrig's men sat the red clay model in the center of the room, with Ames' ideas pushed into the corners. Ames was not pleased.

The interior of Buehrig's Duesenberg Murphy Beverly of 1930, and a similar view of his Cord 810 armchair Beverly interior of 1936. (Don Howell, Don Randall)

Beverly sedans, whose bodies had been built by coachbuilders Murphy and Rollston. The Duesenberg's flat broadcloth panels were adapted for what would become the Westchester interior; the fixed armrests would be the basis for the future 810 Beverly. (Herb Newport suggested some of the interior details. Dome lights for the new sedan were inspired by those on his Walker hardtop coupe, as was the bezel of the top-mounted radio speaker.)

The new car's instrument panel would later be lauded as one of the most impressive ever put into a production car. Buehrig leaned heavily on the most modern icon of the 1930s—the airplane. More instrumentation was provided than was typical. Large round speedometer, tachometer, oil pressure gauge and clock dominated the panel. Temperature gauge and radio dial were smaller circles. Ammeter and gas gauge were fan-shaped. Levers for light switches, hand throttle, choke and panel lights were miniature versions of airliner throttles.

The panel that surrounded the instruments was chromed and "engine-turned." This process, also called "damascening," resulted in hundreds of overlapping shimmering circles, creating multiple reflected holographic points of light.

The dashboard of E306/2. For production a round temperature gauge replaced the rectangular one and the ignition switch was moved to the dashboard as part of the ammeter assembly. The single headlight handle at lower right was for the Ames headlight operating mechanism. (Randy Ema)

All the instruments were stock units from Stewart-Warner and King-Seeley. The latter supplied an older water temperature gauge that was based on a thermometer. For this and all the rest Buehrig designed matching new faces, with some of the graphics engraved on glass. When combined with Magnavox's new concept of edge-lighting, the green glow of the dashboard at night took one's breath away.

The red clay model had a rather vaguely shaped transmission cover. The gearshift mechanism had not yet been designed, and no one was sure yet what its space requirements were going to be.

Buehrig's vision could turn practical engineering into fine art. An example was the design of the doors. The baby Duesenberg used front-opening doors front and rear. (This type of door is called a "suicide door" today, but was not so referred to then.) Buehrig planned to hinge both doors at the center pillar, because it offered another opportunity to save on scarce tooling dollars. The left front and right rear door could be stamped from the same dies; ditto for right front and left rear. A small trim die would create the cutout in the rear door design to clear the rear fender. The intention was economy; the effect was art.

The new car's striking beauty was in its shape, not its chrome decoration. Only bumpers, wheel covers, windshield frames, handles and locks would be brightwork. Initial thinking was that

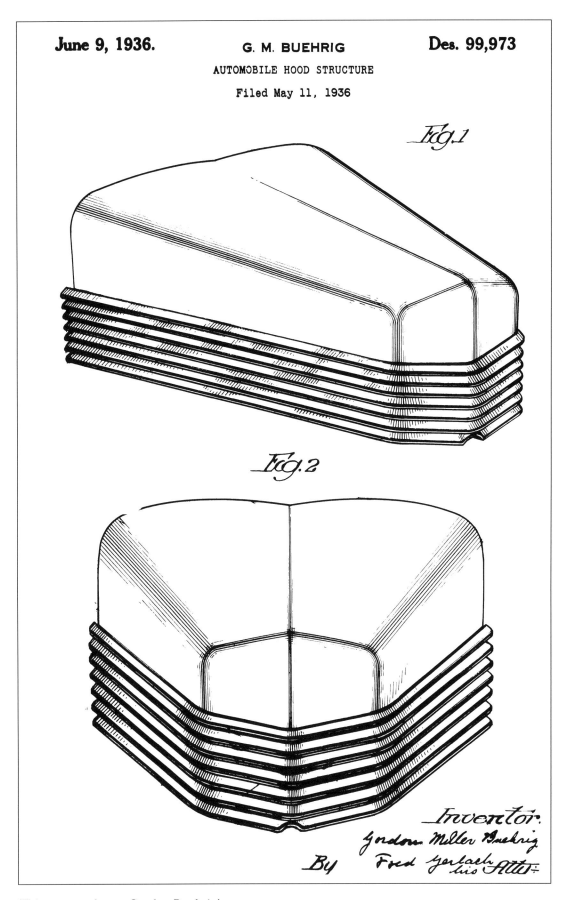

This page and next: Gordon Buehrig's design patents.

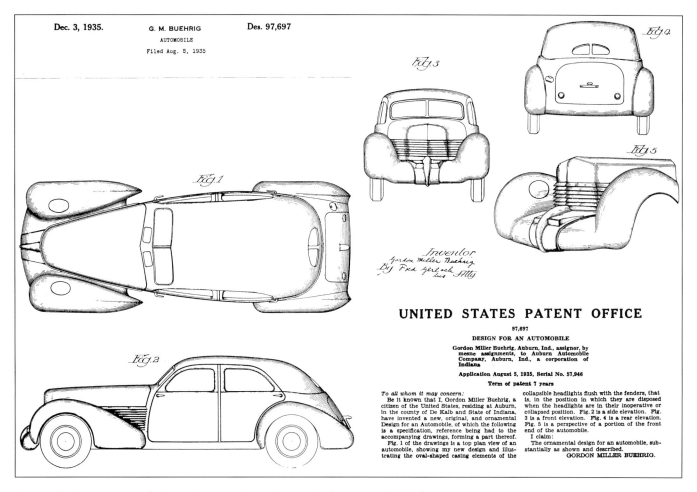

the grille louvers might be chrome-plated or stainless steel. Neither was eventually chosen.

Gordon Buehrig and his design team were a mutual admiration society. He always referred to their partnership whenever he was praised for the design of the Cord 810. They never failed to honor him as the leader of the effort. And, unlike the practice in other design studios, the head of this team was involved in every detail of the job.

Word of the quarter-scale model reached Chicago. The board of directors of the Cord Corporation probably met quarterly or semi-annually, so there's likely to have been a meeting scheduled for early January. In preparation for this meeting, Lou Manning asked Gordon Buehrig in late November to bring the quarter-scale red model to Chicago. Manning was not a car enthusiast; his interest was in the corporation's financial well-being. He and Hube Beal met with Gordon and looked at the model. What they thought and what they said is not recorded.

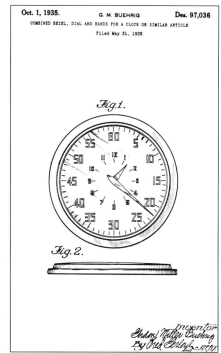

5

POWERED BY LYCOMING

Chroniclers of the development of the Cord 810 usually dwell on the innovative combination transmission and final drive (the term "transaxle" had not yet been coined), and on the necessary constant-velocity universal joints. While the Cord's V-8 engine is often faintly praised, it has not received a great deal of attention.

Sad oversight. The V-8 that Lycoming developed for the Cord is worthy of note. It's a fascinating combination of some surprising anachronisms with the latest in engine design for its time.

Dozens of independent manufacturers provided the motive power for American automobiles from the beginning of automobile production in this country until the 1950s. Few, though, attained the stature and fame of Lycoming Manufacturing Company of Williamsport, Pennsylvania. Lycoming was founded in 1908 as a foundry and machine shop and built its first automobile engine in 1912. In 1923 Lycoming purchased the capital stock of Spencer

An early FB engine. Changes were made before production began. These include the rib pattern on the heads, elimination of the chain case-mounted breather, a two-unit regulator and a larger fuel pump. This obsolete photo was circulated for the life of the car. (Randy Ema)

Heater Company. It provided a then-revolutionary straight-eight engine to Auburn (and Elcar and Apperson) for their 1924 models.

In 1927 E.L. Cord gained a controlling interest in Lycoming, and it became a subsidiary of the Auburn Automobile Company. When the Cord Corporation was formed in 1929, the rest of Lycoming's stock was acquired as well. Lycoming eventually manufactured 57 different automobile engines that powered 250 models of 46 makes of cars, including Auburns, Cords and Duesenbergs. Its marine engines were often used in racing. It supplied nine-cylinder radial airplane engines to Stinson, and horizontally opposed fours, sixes and eights to Piper, Ryan and Beech. (One Lycoming experimental airplane engine, the 36-cylinder 5,000 horsepower XR-7755, remains the largest radial engine ever built.) Lycoming-produced engines were designed by the company's own engineers, with the notable exception of the dual overhead cam Duesenberg Model J.

What the FA engine may have looked like. This reconstruction is based on the FA parts list, the components of the Lycoming BB V-12 engine, the cylinder head patent drawings, and descriptions by August W. Rickenbach. External differences from the FB series include a single pulley driving the water pump and generator, full flow oil filter, timing chain case casting incorporating the water pump, early Fahlman patent heads, single thermostat, and exhaust-heated intake manifold. (Ken Eberts)

Retired Lycoming employees have confirmed that the company did not build automobile engines on speculation, for possible later sale to a car maker. All designs were prepared based on requests by the potential customer. In the fall of 1933 Auburn appears to have requested proposals for a V-8 engine for a rear-drive car. There is no indication in surviving internal memos of any plan to power a future Auburn with a V-8. The request to Lycoming came right around the time that Ames lunched with Gordon Buehrig. Was Ames considering a new Lycoming V-8 as the power plant for his baby Duesenberg?

Forrest S. (Bill) Baster was then chief engineer at Lycoming. He'd arrived in Williamsport in 1932, fresh from success at Hupmobile. Baster's team laid out a 90 degree V-8 to be cast in a single block. Like other Lycoming-designed engines, it received a two-letter designation. "F" was the letter planned for future V-8 designs. So, this first V-8 was labeled "FA."

Specifications for the new FA engine were quite modern. A counterbalanced crankshaft turned in three large main bearings. Cylinder heads were cast aluminum. So were the 4-ring pistons, manufactured for Lycoming by Ray-Day. A bore of 3-3/8 inches and stroke of 3-3/4 inches yielded a displacement of 269.4 cubic inches. Removable side plates provided access for cleaning core sand from the cast block, and to keep water passages clean during service.

Most interesting was the valve train. A single camshaft was mounted high in the block. Cam lobes met a solid rocker midway along its length; the rocker in turn operated the valves. This permitted the valves to lie nearly horizontal. Cylinder head decks

Form ED-272 10M 2-36 AMERICAN FORM			LYCOMING FA SERIES		11-27-33	11 Sheets Sheet No. 7	
ENGINEERING PARTS LIST			3-3/8 X 3-3/4 - 8 CYLINDER REAR DRIVE SAMPLE ENGINE				
Assembly Symbol	Part Symbol	No. Req.	PART NAME	Material	Use Instead of	REMARKS	Wt. per 100 Pcs.
	F-4450	1	Oil Pump Idler Stud				
	WA6	2	Washer - 3/8 Lock				
	8A610F	2	Capscrew - 3/8-16 X 1-1/4 Hex				
	F-4410	1	Oil Pump Cover				
	WA4	4	Washer - 1/4 Lock				
	SG406	4	Screw - 1/4-28 X 3/4 Fill Hd				
	F-4620	1	Oil Strainer Body				
	WB5	2	Washer - 5/16 Lock				
	SZ9502	2	Capscrew - 5/16-18 X 3/4 Hex				
	F-4622	9	Oil Strainer Body Gasket				
	F-6500	1	Oil Filter Base				
	F-6551	1	Oil Filter Base Gasket				
	SA624F	2	Capscrew - 3/8-16 X 3 Hex				
	SA609F	1	Capscrew - 3/8-16 X 1-1/8 Hex				
	G-472	1	Oil Relief Plunger				
	G-473-D	1	Oil Relief Spring				
	H-474	2	Oil Relief Spring Spacer				
	G-474-B	1	Oil Relief Spring Plug				
	1136	1	Gasket - 3/4 X 1 Round	Copper Asbestos			
	E-19519	1	Oil Filter (buy part)				
	E-6501	1	Oil Filter Body Gasket				
	SA608F	3	Capscrew - 3/8-16 X 1 Hex				
	WA6	3	Washer - 3/8 Lock				

Top: A sheet from the parts list for the FA rear drive engine, showing parts for the full-flow oil filter. (Stanley Gilliland)

Bottom: Forrest "Bill" Baster. This photo was taken in the 1950s, when he worked for White Motors. (Owen Obetz)

were not perpendicular to the cylinder bores, and the result was a wedge-shaped combustion chamber, contained entirely within the block. The cylinder head served only as a roof for the combustion chamber. It had a flat working surface, broken only by reliefs for the valves and pistons. This design resulted in relatively high volumetric efficiency for a nominally side-valve engine. It permitted large valves, with little shrouding. Exhaust ports exited at the top of block, so good cooling was expected since the hot exhaust gases didn't pass through the block. Smoothly curved cast headers portended low back pressure.

The FA design included full-flow oil filtration, not unlike the system used on the V-12 which Lycoming had created for Auburn in 1932. Also planned was an oil level gauge, which would permit the driver to determine the oil level from the driver's seat without touching a dipstick.

But good design on paper doesn't always translate smoothly into cast iron and steel. There were some early omens of what was to come. August W. "Rick" Rickenbach headed Lycoming's experimental department from 1932 to 1937. A native of Williamsport, he'd started work with the company as a junior engineer in 1929. He retired from Lycoming in 1965. He remembers that the sample V-8 engines were the only models Lycoming ever built that experienced vapor lock and fuel percolation while being tested on the dynamometer!

Pondering this problem, Rickenbach sketched some alternative layouts on a yellow scratch pad. He replaced the prototype engine's traditional exhaust-heated intake manifold with a design in which heated water flowed through the intake manifold while the engine was warming up. To provide for rapid heat transfer, the new intake manifold would be cast of aluminum. Water flow was to be controlled by the cooling system thermostats. Cylinder heads in the prototype FA had their water outlets at the front. That required different castings for the right and left head. And the FA used a single thermostat, which would have made the plumbing for the new intake manifold design more convoluted than was desirable. Rickenbach sketched a variation with a thermostat housing centered on each head, which would permit the identical head to be used on either bank. As he tells it, "Mr. Baster came by and I showed him my sketches. He said 'Let me have those' and took them upstairs." Rickenbach's changes would soon be incorporated into the design of the new V-8.

The parts list for the FA V-8 sample engine is dated November 27, 1933. That fits the date when Ames approved Buehrig's baby Duesenberg's sketches, and lends strength to the possibility that the V-8 was to power the car. By June 1934 Ames had made the recommendation to not proceed with the baby Duesenberg project. Instead, Auburn would pick up the front wheel drive banner it had laid down in 1932. The revised version of the Lycoming V-8 would be the series FB.

The front drive FB engine required pulleys at both ends of the camshaft. One drove the generator, the other the water pump and fan. Other changes were made after testing of the prototype FA. A cam lobe sliding over a solid rocker generates tremendous localized pressures. The simple oils of the day couldn't handle this, and cam lobes scuffed in tests. To deal with this problem, a roller was designed into each valve rocker to reduce cam lobe wear. (Lycoming had used a similar design in the Auburn V-12.) One odd change was the reduction in the number of piston rings from four to three.

A change which was not made was the location of the oil pump. In most cars the pump pickup is located toward the rear of the oil pan, so on steep hills the inlet will have plenty of oil to draw from. The FB engine has its oil pickup located toward the front of the engine, which was intended to be the rear when the design was first created.

The first sample front drive engine, FB 102, retained the exhaust-heated intake manifold of FA 101. FB 103 was the first engine with Rickenbach's water-heated aluminum manifold, and the first engine delivered to Auburn.

Ventilation of an internal combustion engine's crankcase is necessary to reduce acid fumes which encourage the creation of sludge. Well into the 1950s, most American car engines ventilated the crankcase by means of a road draft tube. Opening into the crankcase and pointed down toward the road, the tube had its bottom end cut at a 45 degree angle. The resulting negative pressure when the car was moving was supposed to draw air through the oil filler cap and out the tube, carrying with it noxious crankcase vapors. The fumes were released into the air; environmental concerns were still many decades in the future.

Road draft systems work only when the car is moving at a substantial speed. Lycoming designed a rudimentary, valveless

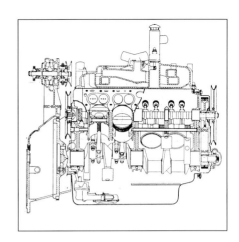

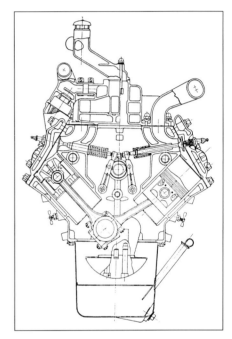

Cross-sections of the production FB engine.

> Just as you get Cream from the top of the bottle—
>
> # FLOAT-O
>
> Supplies bearings with the "Cream" of the Oil—from the top of the Crank Case
>
> The sludge, filings, and heavy abrasives which cause serious engine wear and inefficiency naturally precipitate to the bottom of the crank case. FLOAT-O installed at the pump intake, draws horizontally from the clean oil found at the top—it does not disturb the harmful substances found at the bottom of the crank case. With FLOAT-O only this "cream" of the oil sump is distributed to the bearings. This is true during starting and all running conditions. FLOAT-O is also a definite guarantee against ice locking.
>
> Indorsed and approved by the leading research engineers of the industry, FLOAT-O insures quicker starting, smoother operation, and longer life for engines.
>
> **The following outstanding manufacturers use FLOAT-O**
>
> | Allis Chalmers | Buick | Lycoming Motors | Pierce-Arrow | Willys Overland |
> | Auburn | Cadillac | Morse Motors, Ltd. | Studebaker | Wolseley Motors, |
> | Buda | Int'l Harv. Co. | Otto Engine | White Motors | Ltd. |
>
> Two other prominent builders definitely committed for 1938 models.
>
> FLOAT-O Engineers are ready to consult with you.
>
> **WRITE FOR LITERATURE**
>
> # TAYLOR
> SALES ENGINEERING CORP.
> Elkhart, Indiana
>
>
>
> *February 27, 1937*

An ad for the Float-O oil pickup. It was used on the Cord as a substitute for oil filtration. (Ron Irwin)

positive crankcase ventilation system for the FB engine. A 1/2 inch copper pipe ran from the cast aluminum oil filler to a hole in the air cleaner. Air flowing into the carburetor throat would create low pressure at the top of the pipe, drawing oily vapors from the crankcase to be burned in the cylinders. Incidental (and accidental) environmental benefits were produced. Clean replacement air was drawn in through a filtered breather on top of the timing chain case.

The FB engine was fully pressure lubricated. Main and rod bearings received oil under pressure. A tiny passage in each valve rocker provided oil to the pins on which the bronze-bushed steel rollers turned. A port even sprayed lubricant on the timing gear and chain.

Despite this careful attention to lubrication, full-flow oil filtration was dropped from the final design. The oil filter on the rear drive FA engine would have been located at the right rear of the block. That's where the oil pump and main oil galleys were. When the block was turned end for end to create the front drive FB, the filter would have moved to the left front of the car. There it would have interfered with the steering box and steering column. (This problem did not occur on the V-12. Rickenbach points out that the Auburn engine was a 60 degree V design, making it quite narrow. That left plenty of clearance for the oil filter.) It appears that Lycoming next proposed bypass oil filtration as a substitute. A bypass filter could be mounted on the firewall, solving any clearance or accessibility problems. An oil outlet from the engine block would be needed. So Lycoming cast a boss into the engine block, just below the outlet for the oil pressure gauge, to be drilled and tapped for the filter outlet. This boss, undrilled, is found on the first 100 engine blocks. Later FBs don't have this boss, nor do they have any oil filtration at all. It's possible

> ### First Come, First Served
>
> Truth be told, Lycoming's series designations were often inconsistent. Lycoming employee Jim Zerfing has tried to reconcile thirty years of Lycoming series letters, with inconclusive results. Other company sources confirm that "F" usually meant V-8, but not always. One former employee told me that a given letter designation went to "whoever grabbed it first"!

that cost considerations influenced the final decision. As a concession to the need for clean oil, the "Float-O" oil pickup was incorporated into the design.

Main bearings on the new engine are described as "semi-finished." This means that relatively thick bronze half-shells with babbitt surfaces are fitted to matching cavities in the engine. After the bearing caps are installed, the bearings are line-bored to their final size and finish. This was midway technology between the poured-in-place babbitt bearings of earlier years, and the precision thinwall shells that were to come later. Rod bearings were of babbitt spun into the rod. Short-sighted was the decision to provide no take-up, manual or automatic, for timing chain wear.

Auburn and Lycoming considered three different sizes of FB engines. All had the identical 3-3/4 inch stroke, like the FA. Bore varied; strangely, none of the FB variants used the 3-3/8 inch bore of the FA. The 3-1/4 inch bore FB displaced 249 cubic inches, the 3-5/16 inch version 258.5 inches. The final 288.6 cubic inch production engine had a 3-1/2 inch bore. That puts it within a fraction of an inch of being "square," defined as bore and stroke of the same dimension. Since piston speed bears a rough correlation to engine revving capability and to potential longevity, short stroke engines are considered desirable. Square and "oversquare" designs, the latter with bore larger than the stroke, would not be seen in American production engines until the overhead valve V-8s of the 1949 Oldsmobiles and Cadillacs.

Lycoming had no facilities for casting aluminum, so fabrication of such parts was contracted to outside suppliers. The Permold Company of Cleveland, Ohio was one of these. They provided a

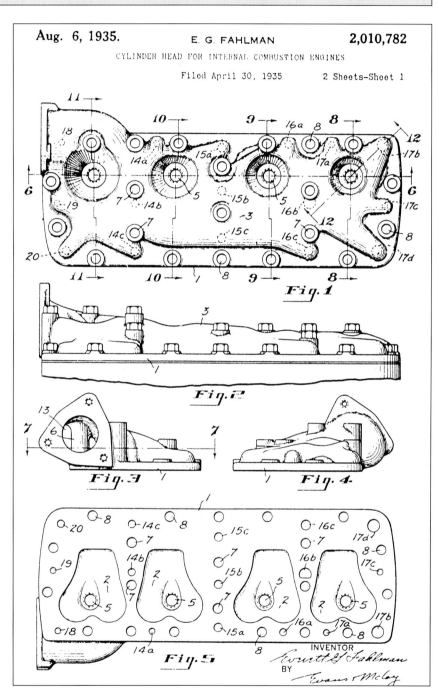

Patent drawings for the Fahlman head.

Technical: Cylinder Heads

Cylinder heads on typical flathead engines were just that—flat. Aftermarket high compression heads were often decorated with fins, but they were still essentially uniform in cross section. The heads on the Cord V-8 have interesting shapes embossed on the surface and vary in thickness from end to end and from top to bottom. Here's why.

The Permold Company, one of Lycoming's suppliers, had been researching the problem created by the varying temperatures of combustion chambers caused by the pattern of water flow within the cylinder head. In a typical head of the time, water entering from the engine block flowed in a substantially random pattern. Streams from several inlet holes commingled and disrupted each other's flow. Although each combustion chamber was bathed by the same quantity of water, the coolant became hotter as it moved toward the outlet from the head, and so was less able to absorb the heat of combustion. The result was head temperatures that varied over the length of the head. The problem was exacerbated by aluminum alloy heads, with their higher thermal conductivity. Under heavy load, uneven running and potential head cracking were some of the consequences.

Everett G. Fahlman, a Permold engineer, had been working with a unique head design since the early thirties. He approached the problems from several points. First, internal baffles guided the entering streams of water so they played over separate combustion chambers with minimum turbulence. Second, the height of the coolant passages in the head increased as they approached the outlet port. The water passing over the combustion chambers closest to the outlet may have been hotter, but there was more of it, in compensation.

In 1933 Permold offered this design to Lycoming for use on the FA V-8 engine. Fahlman applied for a patent in April 1935. It was granted on August 6, 1935, and assigned to the Permold Company. The Cord V-8 was the first engine to use the new design.

The FA engine had each cylinder head water outlet at the front of the car, closest to the radiator. Fahlman's patent drawings were probably made from this prototype head. When Lycoming's Rickenbach proposed a centered outlet, Permold adapted the Fahlman principle, so coolant entering from both ends of the block flowed toward the middle.

Examination of a factory-supplied Cord cylinder head shows that in cross-section each head is thinnest at the edges. Larger passages cause the head to grow deeper as the coolant approaches the top center of the head, where the water outlet is located. As in Fahlman's drawings, the shape of the coolant passages is visible on the outside of the head. The varying depths mandated by the Fahlman design is the reason that three different lengths of head bolts are needed on the Cord engine.

Lycoming made several recommendations for changes in the ribbing and internal construction of the cylinder head, mostly for reasons of strength. The final Permold design was used on all production engines. One can identify the early factory-supplied Cord cylinder heads—stamped on the outside of each is Fahlman's patent number.

cylinder head of patented design, intended to permit more uniform cooling. Lycoming took other pains with the design of the FB's cooling system. A powerful water pump with dual chambers pumped coolant into both sides of the block. Brass water distribution tubes brought coolant directly to the valve seat areas along the full length of the block.

Lycoming horsepower and torque curves, dated May 1934, show output from both normally aspirated and centrifugally supercharged versions of each engine size. The legend on the sheets suggests that the charted outputs were theoretical, based on the performance of other Lycoming engines, and that no actual experimental engines were ever built in the smaller sizes. The FB engine did, however, provide from the start for the possibility of the later addition of a supercharger. The camshaft had a shoulder on which a bevel drive gear could be installed, and the necessary mounting bosses were cast in place in the block.

Time has demonstrated the quality of Lycoming's engineering and production abilities. The FB and later FC engines proved reliable and long-lived. In evaluating them remember that Ford's 1932 V-8, released only two years before the Cord engine was designed, was revolutionary in bringing V-8 power to a low-priced car. Ford was able to accomplish this by innovative casting technology, and the anticipation of millions of blocks over which to amortize the expensive tooling. The only other production American V-8 in 1936 was Cadillac, whose output for two months exceeded Auburn's total annual projections for the Cord. Then came Lycoming, tiny by comparison with Ford, with the grand idea of building a V-8 engine for a car whose most optimistic production plans measured in the thousands. Auburn presented Lycoming with a daunting challenge, and the Williamsport company met it bravely.

By the end of 1934, Lycoming was ready to deliver FB engines to Auburn. We don't know why Auburn made the decision to delay introduction of the supercharged engine to the 1937 model year. It may have been because they wanted to save another bombshell for the 1937 show, or—more likely—because they felt that they already had a bit too much on their plate.

6

YEAR'S END

Auburn ordered up tooling estimates for E278 in November 1934. Stanley Menton, Production Body Engineer of Central Manufacturing Company, prepared the estimate. He wrote, "Body and chassis frame is [sic] to be an integral unit to front of dash.... Fenders and chassis sheet metal are to be similar to sample automobile which writer saw at Auburn," apparently referring to the red quarter-scale model. No working drawings were made available. Menton indicates that all he had to work from was a quarter-scale sketch and verbal information. "The writer had to use his imagination on details." One of those details was the roof panel; absent other information, Menton imagined composite construction like an Auburn, rather than all steel. Parts to be struck from the dies spelled out in the tooling list fit the description of the Buehrig design's body panels. The list referred to a "V" windshield, as well as trim dies for making rear doors out of front door panels.

Skilled workmen craft the die model for the Cord 810 sedan. When complete this wood car will be cut apart so permanent dies for the body panels can be cast from the pieces. Imperfections in the surfaces of this model will show up in the dies, so a very smooth finish is required. Note that the man at the rear is actually reflected in the wood trunk lid. Dies will be made only for the front doors, since the rear doors are mirror images. Visible on the right front door is a scribe line showing the cutout for the left rear door. (Henry Blommel)

59

Auburn's administration building and main showroom on South Wayne Avenue in Auburn. The photograph was probably made in about 1934. (Ron Irwin)

Doors were to be hinged at the center pillar. There was also a "louver assembly," and fenders made out of several pieces. (Interestingly, the transmission cover was referred to as a "scoop," the slang many Cord owners use for it today.)

Cost estimates were worked out for six and V-8 versions. Unit cost for the six cylinder chassis were estimated at $322.90. The eight cylinder version would have cost $407.29. Adding in the identical unit body for both ($271.00), direct charged labor ($41.00) and freight costs brought the total to $656.04 for the six and $774.39 for the eight.

Whether powered by Continental or by Lycoming, there is no evidence that an E278 car was ever built. The last report on this project is in December 1934, when management was offered several alternatives for how tooling up for this new car might be accomplished. One possibility was to essentially build the bodies at Limousine Manufacturing Company, using power hammers and forms. (Formerly of Kalamazoo, Michigan, Limousine Manufacturing had been purchased by Cord and its facilities moved to Connersville. It did specials and custom work for Auburn.) Another

was to contract the work out to Budd Manufacturing Company. This document, recorded on Friday, December 21, 1934, is the last one to refer to E278.

The basis for the red clay model was project 286, an 8 cylinder special front drive. At 123 inches the wheelbase of E286 was three inches longer than E278, but it retained E278's four wheel independent suspension. Buehrig's drawing, dated September 28, 1934, is the first known record mentioning E286. Graphs prepared on December 20 have to do with wheel angles and turning geometry. These were undoubtedly needed to examine the effects of the longer wheelbase.

When 1934 ended, no front wheel drive project car had yet been built. Each of the new project numbers represented the next stage of the same design, although the new stage was usually begun before all possible data had been extracted from the previous one. E278 had a 120 inch wheelbase, a three speed transmission and four-wheel independent suspension. It had a unit body from the cowl back, based on Buehrig's design. Calculations, both engineering and financial, were done to enable the choice between a Continental six and a new Lycoming V-8. (If the V-8 were chosen, further choices would then have to be made regarding displacement and supercharging.) E286 was essentially the same car with the wheelbase lengthened to 123 inches, and the Lycoming FB engine of 288 cubic inches as the powerplant.

Project E294, with a wheelbase of 125 inches, would be the next step. While final chassis engineering was still going on, management approved the body design. Tooling began.

During the fall of 1934 Gordon was courting Betty Whitten of Auburn, and on December 22 they were married. They honeymooned in Fort Lauderdale, Florida. When the Buehrigs left Auburn, full size body drafts had been prepared for the sedan body of the new front drive car. Wooden die models for this sedan body were well along in construction. The rest of the Body & Art department was working on a quarter-scale clay model of a convertible quarter window victoria.

Optimism was in the air in Auburn during those closing days of 1934. The next big auto shows were scheduled for January 1936, more than a year away. Development work for what would be Auburn's latest blockbuster was right on schedule.

7

1935

One of the persistent legends in the history of the Cord deals with the project's death and resurrection. The story goes that Cord Corporation management in Chicago killed the project at the end of 1934. (Gordon and Betty Buehrig were honeymooning then.) The story then dwells on a striking series of photographs that were shot only one day before they were presented to a July 1935 meeting of the board of directors of the Cord Corporation. Impressed, the board approved the project six months after it had originally shelved it. Gordon and others have referred to this seemingly senseless delay of six months as a major reason for the demise of the Cord two years later.

Using the additional knowledge we have today, and some educated guesswork, the story comes out a bit differently.

Meetings of the board of directors of the Cord Corporation were likely to have been scheduled around the end of each quarter of the fiscal year. There are no surviving minutes. We can only guess at

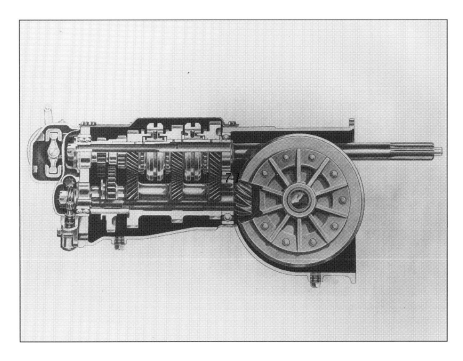

Lorenzen's sectional view of the Cord drive unit. The early Archimedes screw-type pump is shown. (Randy Ema)

63

Bendix heavily advertised its "Finger-Tip Gearshift" to the auto manufacturers. Only two ordered it: Hudson (for the Hudson and Terraplane) and Auburn (for the Cord). (Ron Irwin)

what went on by the decisions they announced. Lou Manning and Hube Beal had looked at the quarter-scale red model that Buehrig brought to Chicago on November 28, 1934. Decisions on the nature of the 1936 Auburn models were on the next board agenda. The board took action. Auburn was ordered to wind down the project, which had been to create a totally new car. The reason was simple—the company had run out of cash. Auburn's books closed in December 1934 with a net loss for the year of $3,900,000. There just wasn't enough money to create the tooling and marketing campaign necessary to launch a brand new make, no matter how high the hopes for it might have been.

Buehrig returned to work after his honeymoon on Monday, January 7. He found that the decision of the board of directors of the Cord Corporation had been announced at Auburn. The proposed program for a brand new car, they said, was too expensive. Instead, they decided the company should work something out for the 1936 models using as much existing tooling as possible. The board had not said, however, that the front drive car could not be in Auburn's 1936 line.

Herb Snow was firmly in favor of a cheaper compromise program that would meet the board's dictates. Bolt E286's chassis to an Auburn frame, and mount an Auburn body. New front end sheet metal like the red model would give the result a whole new look. A whole new ugly look, thought Buehrig and his team, but they loyally produced a new quarter-scale model to prove it.

Engineering studies of project E294, the next step in front drive development, had already begun by the time the corporate board acted. Theoretical acceleration and speed studies were based on the middle-sized FB engine, and a three-speed transmission. Comparison curves done in January and early February compare the new design's projected performance against an Auburn 851. For purposes of calculation, frontal area was estimated as the same as the Auburn, minus a proportional reduction for 3 inches less height. The performance comparisons presented the engineers with a dilemma. By properly using its two-speed rear axle, the production 851 was capable of comfortably out-accelerating the new car, then adding to the insult by reaching a higher top speed.

That just wouldn't do; design changes were obviously needed. Weaver would no doubt be ordered back to work with Detroit Gear and Machine to add a fourth speed to the transmission. Lycoming would be informed that the new car would be powered by the largest of the FB options, and instructed to prepare sample engines.

Snow was serious about rescuing the front wheel drive design into which he had poured so many man-hours and so much of himself. His hybrid E294 was a safe bet. If the Buehrig-bodied new car were ever to be approved, E294 was a useful test bed for the mechanical design. If not, perhaps E294 as built could be the prototype for an Auburn line for 1936 that would meet the corporate board's demands for economy.

64

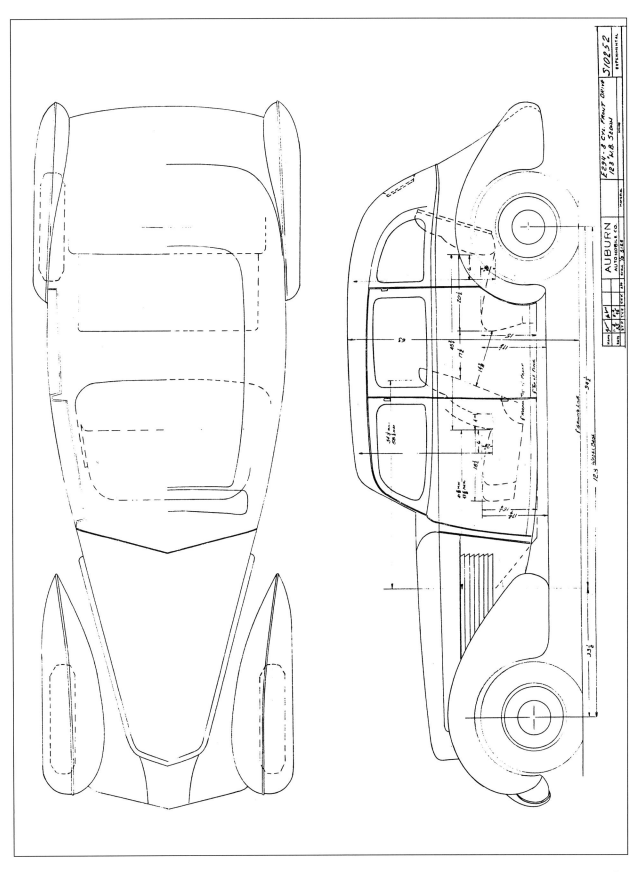

This drawing of E294 was done on January 4, 1935, by Stanley Thomas, Auburn's development engineer. It shows the layout of the front drive hybrid ordered by Snow immediately after the board decision. Buehrig had not yet returned from his honeymoon. (Redrawn by Robert Fabris)

Artist's conception of the E294 "mule," based on Thomas' drawing, and on the engineering studies and test reports. The car was nearly three inches taller than the production Cord, but still three inches lower than a stock Auburn. Note the moldings on the fender. (Ken Eberts)

And so the engineers were to build, drive and test Snow's front drive Auburn hybrid. Since it was the first running prototype of their new front drive design, they naturally called it E294. Wheelbase was determined by the length of the front drive stub frame forward of the cowl, and the length of the Auburn body from the cowl back. It wound up as 128 inches.

It's important to distinguish between **project** E294 and **car** E294. The project was the design of a front wheel drive sedan with body based on the red clay model, a further development of projects E278 and E286. The car was the front drive mule with the front sheet metal of the clay model mated to an Auburn body. When actual prototypes were later built from project E294 they were called project E306. To compound the confusion, the engineers stubbornly continued to use the terms E294 and E306 interchangeably as late as August 1935!

E294 was the most important project under way at Auburn during the months of January, February and March 1935. Personnel records show this. Work moved ahead on two fronts. The Auburn-bodied hybrid car was being built in the Experimental Wood and Metal Shop and the Experimental Garage. Simultaneously, work on project E294, the design based entirely on the red quarter-scale sedan model, continued on the drawing boards upstairs. Factory records show that 1412 hours were spent by personnel in Body & Art, Chassis Drafting and Design & Research departments on project E294 in January 1935.

The decision to provide four speeds in the transmission made the gearshift mechanism more complicated. Bragg-Kliesrath Division of Bendix Aviation Corporation had demonstrated its seductive new "Finger-Tip Gear Control" to manufacturers in 1934. A gear selector switch mounted on the steering column mimicked the pattern of a floor shift, but with a lever small enough to be

The Gentlemen's Speedster, an attempt to use up parts from the Cord Corporation's three cars. Only one was built. It was never formally associated with any of the makes. (Auburn-Cord-Duesenberg Museum)

moved by the driver's finger. Hudsons and Terraplanes offered the device on their 1935 models. Hudson called it "Electric Hand."

Auburn engineers explored alternate ideas for mechanical shift linkages. An unfinished drawing shows a shift mechanism built into the left side of the transmission, with a linkage of parallel rods operated by a column-mounted shift lever. A decision had to be made. The Bendix gearshift had in its favor the ease of connecting the steering column shift with the mechanism up front. Wires are easier to route than rods and bellcranks. An ancillary advantage was the very modern look that the miniature selector gave to the car's interior. On the other hand, the mindless vacuum cylinders slammed gears into synchronization, often with a palpable clunk. This boded ill for gear longevity.

Expediency prevailed. The Bendix unit worked, it was available, and Auburn had no time to fuss with this detail. (Not all of Auburn's engineers were convinced. At least one drawing of a mechanical shift is dated as late as April 1935!)

The Cord gearshift is the only Finger-Tip Gear Control ever supplied by Bendix that was designed as an integral part of the transmission, and the only one that selected four speeds. The components on the Hudson and Terraplane were all external add-ons. Indeed, a floor-mounted gearshift lever could be installed to cope with emergency shifting problems.

Buehrig's crew had completed the drawings for the body and interior of the unadulterated project E294 at the end of 1934. Only details remained. With the transmission mechanism determined, Art & Body Drafting could now design the transmission cover. Buehrig says in his book Rolling Sculpture that he ". . . allowed the problem to solve itself." The Body & Art men built a wooden mockup of the right side of the Bendix gearshift, where the mechanical parts projected most. They laid slabs of hot clay one

E306 #2, the first completed Cord prototype. This front view was retouched later for use in advertising. The grooves have been removed from the bumper, and a transmission emblem added. The very low windshield and rear window are obvious in this view. So is the large diameter of the inboard headlight lenses; a different supplier was chosen for the production cars. (Auburn-Cord-Duesenberg Museum)

inch thick over the wood, and smoothed off the lines. Then they transferred this design to the other side to make it symmetrical. Form followed function.

Other changes from clay model to prototype design were related to the state of the art in 1935. The clay model's one-piece rear window would have required curved glass. This wasn't available then at a reasonable price, so two small flat panes were used in the final design. And since the front fenders could now be made without a visible seam, the chrome center molding was dropped.

Far from being a period of suspended animation at Auburn, the months of January, February and March 1935 were filled with activity. Probably the only quiet place was Buehrig's Art & Body Drafting shop. Until a decision was made about going forward with one car design or another, they had time on their hands. So they collaborated with other departments in the creation of a prototype two-seater Gentleman's Speedster. An Ames idea, this creation was to use up leftover Auburn 12-cylinder engines and possibly create an alternative cheaper Duesenberg. At least one prototype was built, but the concept went no further.

The relatively quiet period permitted the Body Art & Drafting team to work on another project requested by management. Tooling and unit production cost estimates for the E294 car had been completed. Clearly, the new front drive car was never going to be transportation for the masses. But its sporty, close coupled design might not appeal to the conservative customer of means, either. What might work was a special custom series. The same stunning lines as the new car, but with more room for the wealthy passengers to stretch their legs. Buehrig and his crew were instructed to draw up such a body design.

The other offices and engineering workrooms at Auburn hummed with activity in those early months of 1935. Engineers manned their blueprint machines, purchasing agents and secretaries their telephones and typewriters. Plans were sent to vendors, and requests for proposals solicited and approved. The initial orders were probably for parts sufficient to complete five sample cars. The possibility of a major production run was held out; vendors to the automobile industry understood the risks, and responded. Parts were received from makers of electrical equipment, steering gears and bumpers. Patterns were prepared for castings for such large pieces as bell housing and transmission parts. Suppliers created tooling for die cast dashboard and interior parts. For wheels and hub caps. For forged steering and suspension components.

The completed E294 hybrid car used many of these parts. Performance tests began at the end of March. The car was powered by a 288.6 cubic inch Lycoming V-8, number FB103. This was the second FB engine, and the first to carry Rickenbach's water-heated intake manifold, according to the records of the tests. The car's four-speed drive unit was the first of five experimental transmissions hastily assembled by Columbia Axle. The independent rear suspension of the earlier designs had been replaced by a tubular dead axle carried on conventional semi-elliptic springs. (Because the rear springs only carried a lightweight tubular axle, this arrangement met Snow's standards for low unsprung weight. In a comparison with the current Auburn Model 851, it was found that unsprung weight had been reduced by 34%.)

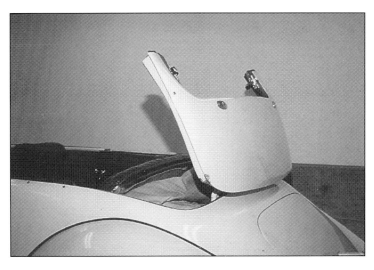

The cloth tops on the convertible coupe and the phaeton disappeared under a smooth steel lid.

In April 1935, at the beginning of spring in Auburn, any semblance of a quiet period was rudely shattered. Factory records show extensive work in the Experimental Wood and Metal Shop on wood bucks for the body and fenders of five hand-crafted prototypes. Body & Art, Chassis Drafting and Design & Research went on full-time schedules. By May all departments were on overtime. They were hard at work on project E306. It was titled "Sedan bodies experimental—build 5." The prototype Cords were under construction.

Frames arrived at the Auburn plant from Midland Steel. Columbia Axle assembled suspension parts. Bendix and Gear Grinding delivered universal joints. The deafening noise of power hammers told everyone within earshot that fenders and doors and bodies were being created by talented artisans.

In that spring of 1935 Columbia axle built five experimental drive units from a mix of temporary and permanent tooling and patterns. The heavy housing for these transmission-differential units was cast and machined by Columbia in Toledo. From Detroit Gear & Machine Works in Detroit came gears and shafts. Bearings and other parts were supplied by outside vendors. Columbia assembled the drive units, and shipped them to Auburn. Here they were mated to the engines and bell housings which had arrived from Williamsport, and installed in the E306 prototypes. (The first transmission went into the E294 hybrid. It was removed after testing was complete and installed in E306 #4).

The E306 prototypes weighed nearly 4,000 pounds. No one appears to have noticed that the brakes delivered by Bendix were of the same size as those originally specified for the 2,900 pound E278. The results were to be evident soon enough.

Buehrig's men were working nine-hour days, six days a week. Their new tasks included final design and body engineering for two open models: a two-passenger convertible coupe with rumble seat, and a four-passenger convertible victoria. This latter model was to include another first. Convertible tops on similar production cars of the era had blind cloth rear quarter panels. Tops folded into a space behind the rear seat, where cloth and bows lay exposed, or

The four-passenger phaeton. The quarter windows in the convertible top were a first. They were dropped down manually, not cranked down like later cars. (Mrs. Errett Lobban [Virginia] Cord)

were covered with a cloth snap-on panel. Buehrig's top had framed glass quarter windows that could be lowered manually into a pocket on each side. The top itself disappeared beneath a steel panel. When the new cars were open, no canvas was to be seen.

Before the mid-1930s, most steering wheels had solid spokes, cast as a unit with the rim. In 1935 Ford Motor Company offered, as an option, a steering wheel with spokes made of stainless steel strands, held together by small clips. The effect was that of the strings of a banjo, and popular usage quickly applied that name to this design. For 1936, nearly every make would offer a banjo wheel on its highest-priced lines, at least as an option.

Steering wheels were made by casting a material around a steel armature. Hard rubber had been the most popular material. Eastman Kodak held the patents on a cellulose acetate plastic that it called Tenite. Small parts like interior knobs had been made of Tenite since 1932, and Eastman now pushed to expand Tenite's use to steering wheels. For 1936, some cars would still use hard rubber rims, but many would be made out of the new Tenite.

Much of the cost to a manufacturer in creating a new steering wheel design was in the tooling for the armature and for the compression molds required to cast the rim and spokes. There was no room in the budget of the Cord project for steering wheel tooling. Steering wheels for the E306s were purchased from stock, and differed from car to car. For possible production, Buehrig had another inspired idea. He visited Sheller Manufacturing Company in Portland, Indiana, the maker of steering wheels for most makes of cars. Because of the recent popularity of the new banjo wheels,

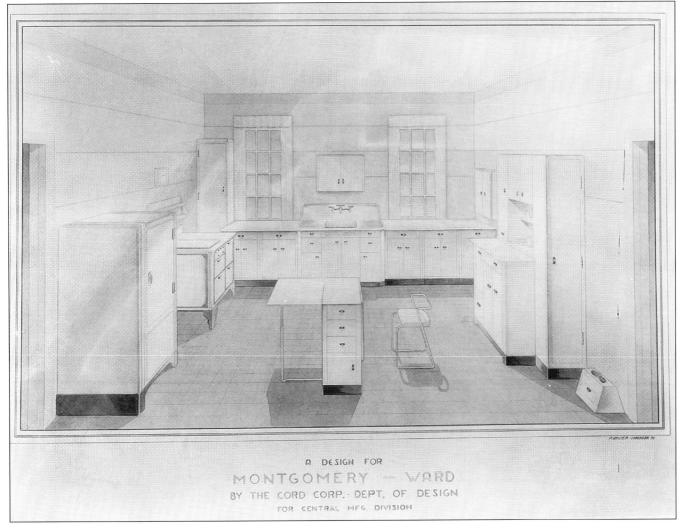

Roy Weisheit's kitchen cabinets; art by Lorenzen. (Henry Blommel)

Sheller found itself with unused tooling for wheels with solid spokes. Buehrig selected a simple design, which Sheller cast in Tenite. Auburn paid only for the production cost of the wheel; there was no tooling cost.

The Auburn design team dolled up this simple wheel with a handsome brass center. Horn buttons on cars of the day were located at the center of the wheel. Buehrig designed a chrome ring that could be used to blow the horn without removing the hands from the wheel. Engineering worked out the electrical arrangement for this first in American cars.

Tooling costs were avoided on the interior hardware too. Buehrig chose an obsolete design from Doehler Jarvis Company in Toledo. Eastman sales representative John Slater gave him a bar of Tenite, from which Buehrig had the shop turn large new handles for the window cranks. With the bold knobs in place, few noticed the older chrome crank handles.

By June the styling detail work was completed, and the pace dropped a bit in Body & Art. It picked up dramatically in all the other areas. Auburn employees worked over 12,000 hours on the Cord project in June alone!

Before the month was over, at least two of the five E306 sedans— the true prototype Cords—were complete. The other three were

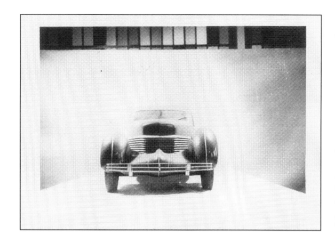

More of Dale Cosper's contact prints, close to their actual size. Only one set was made, so the board members of the Cord Corporation presumably passed these little snapshots around the table on July 8. (Mrs. Dale [Sally] Cosper)

nearly so, and would be trimmed and painted within weeks. The colors were Palm Beach Tan, Clay Rust, Fawn Beige, Black and Cadet Grey—they were, no doubt, a sight to make every employee's heart beat a bit faster.

The five E306 cars were labeled #2 through #6. That's because E294, the hybrid Cord-Auburn, was car #1. From this time forward, the E306s would be the cars on which all testing, design work and tooling and production cost estimates were based, even though engineering and other reports still sometimes refer to them as "E294." (Don Mates calls this "muddying the footprints of history"!)

Was the Cord project actually revived before the July board meeting? By whose authority? Where did the money come from? If the prototypes were indeed completed in June, of what significance were the famous July photos and the Cord Corporation board meeting? And if pictures were needed, why photograph a model when the real thing was standing only a few yards away?

The answers lie, I believe, in actions taken in Washington, DC, in the faith of Roy Faulkner and Harold Ames, in the brilliance of a production man named Roy Weisheit, and in an end run around corporation mandates.

Weisheit was one of Auburn's top production men, responsible for tooling fixtures at Connersville's Central Manufacturing Company plant. It was at "The Central," as it is still called locally, that Auburns had been produced since 1929. As the automotive lines continued to hemorrhage money through 1933, Weisheit sought other products that could be produced in Connersville's partially idle factories. He developed a line of steel kitchen cabinets that could be manufactured on Auburn's stamping presses and welding fixtures. Styled by Paul Reuter-Lorenzen, the modern cabinets appealed to buyers at Montgomery Ward. By June 1935 Weisheit was to close a million-dollar sale, and Auburn would have some of the cash it needed for the Cord project. The rest would come from a stock offering.

Roy Faulkner, president of Auburn, had long been an admirer of Gordon Buehrig's work, and a partisan of the design now under construction. While Executive Vice-President Harold Ames was less impressed, the tricky body that he had first conceived as a small Duesenberg had, after all, been one of his own ideas. Despite the celebrated antagonism between Faulkner and Ames, the potential of the Buehrig-styled car was one idea on which both could agree. I think that Faulkner, aware of Weisheit's kitchen cabinet enterprise, was willing to wager that the cash would be available soon to build the futuristic front wheel drive car. Ames would have reluctantly gone along.

What forced their hand, I believe, was president Franklin Roosevelt's request to the Automobile Manufacturer's Association. He asked that the regional automobile shows usually held in mid-January be moved forward to the beginning of November. On March

Roy H. Faulkner

By the time he was 36 Roy Faulkner had done a stint as county recorder in Pittsburgh, worked as a salesman for a lumber dealer and sold Oaklands and Reos and Stutzes. In 1922 he moved to Cincinnati to become sales-manager of a Nash dealership.

In the fall of 1922 Faulkner was appointed sales manager of the Auburn Automobile Company by the financiers who then owned the company. In November 1925 E.L. Cord became president of Auburn. The new executive team included Faulkner as vice president of sales. When the Cord Corporation was established in 1929, Faulkner was one of the directors. In early 1931, E.L. Cord was elevated to chairman of the board by Auburn's directors, and Roy Faulkner became president of the Auburn Automobile Company.

Nine months later, Faulkner left Auburn to take the position of sales manager for Pierce-Arrow. Auburn was then at the peak of its earnings, but Faulkner was reportedly unhappy with the way the board of directors had handled the distribution of profits. Faulkner would soon became a vice-president of Pierce-Arrow, and president of a national sales group selling Studebakers and Pierces.

In August 1934 the Cord Corporation re-hired Faulkner as president of Auburn. The trade publications were surprised. Faulkner's long-time rival, Auburn executive vice-president Harold Ames, was disgusted. The town of Auburn, however, was ecstatic. Brass bands played, and stores offered special sales. Roy Faulkner was a favorite son, on a first-name basis with most citizens. When he was president of the company, the town was prosperous. After he left, things went downhill. Even the car factory had departed for Connersville. Now Faulkner was back, and happy days were here again.

Roy Faulkner was as low-pressure a salesman as Harold Ames was a hard-driving one. Perhaps Auburn needed stronger leadership than he was able to give it. But it was Faulkner who championed the red clay model of Gordon Buehrig as the basis for Auburn's new front wheel drive car. For this, if for nothing else, he shares in the honors as one of the car's godparents.

Faulkner remained with Auburn through 1939, when it was reorganized as Auburn-Central. During and after W.W.II he carried on a manufacturing business of his own in Auburn. In 1947 he served on the board of the Bobbi Motor Car Corporation. The company was hoping to manufacture one of the many small cars that flickered across the American motor car firmament in those years.

In 1956 Roy Faulkner was assisting the Auburn-Cord Duesenberg Club with the planning of a reunion of car owners in Auburn when he died in Fort Wayne, Indiana, at the age of 69.

15, 1935, the AMA agreed. Auburn no longer had ten months until the 1936 show; they had only seven and one-half months. Something had to happen, and now. To make it happen, they needed the nod from E.L. Cord.

Cord trusted Faulkner. He had just brought him back as Auburn's president, probably over the objections of some of his executives. It is not inconceivable that Faulkner called his boss in March, right after the AMA announcement. He would have told him of the Weisheit project, and suggested that the company get a running start on the front drive project pending the completion of the Montgomery Ward deal. And since it was an open secret that another Auburn front wheel drive car would be named "Cord," E.L. probably approved. Errett Cord always loved a good gamble, and this one had his name on it.

As President of the company, Faulkner had the authority to authorize the resumption of design work, and even the construction of prototypes. What he did not have was the final authority to actually put the new car into production. For this he needed the approval of the board of directors of the Cord Corporation.

So Faulkner and Ames looked for a way to convince Cord Corporation management to bless their project at the next board meeting in Chicago. They couldn't very well present photos of the completed E306 cars, since the development of this project was supposed to have been arrested December last, by order of this very board of directors! The board meeting was set for July 8. On Sunday July 7, Faulkner asked Buehrig for photos of the quarter-

Roy Faulkner. (Auburn-Cord-Duesenberg Museum)

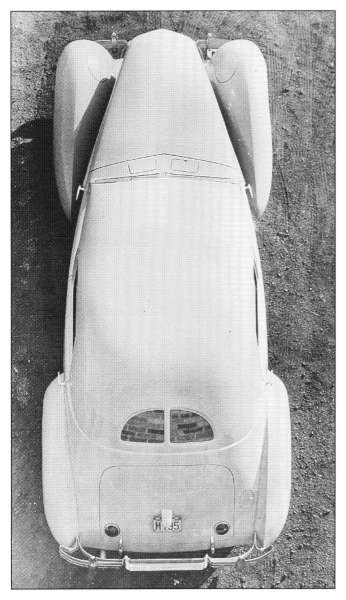 

This car may have been E306 #3, likely the last car built with inboard headlights. It was posed in the factory's gravel courtyard, probably in July 1935. For the overhead shots, the photographer was standing on the roof of the two-story factory building. (Auburn-Cord-Duesenberg Museum)

scale model that the Body & Art team had created in 1934. Buehrig recruited Dale Cosper, who was an amateur photographer. Both of them knew that five full-size cars stood just one floor below in the same building, and must have shrugged at the strange behavior of management types. Together they photographed the model using Cosper's German Steinheil camera. Cosper developed the film and made the 2 1/4 X 3 1/4 contact prints in his own basement darkroom. (Buehrig later said that the prints were made from wet negatives by a process that Cosper had read about in a detective novel!) Barely dry, the prints were rushed to the nearest railroad station. That would have been at Garrett, Indiana, four miles from Auburn. They must have been placed in the mail sack, and the Railway Post Office car of a Chicago-bound passenger train would probably have picked them up on the fly in the middle of the night.

While one million dollars was not much money with which to launch a totally new automotive venture, in 1935 it was still a goodly sum to risk. The fact that an unsentimental board of directors authorized production based solely on a half-dozen snapshots seems to support the theory that their agreement was essentially pro

forma. And that would only have been the case if the chairman of the board in faraway California had already signaled his approval.

On July 29, the first of the five E306 prototypes left on its shakedown run to Los Angeles, so E.L. could see the completed car for the first time. The fabled six month delay never really happened. The folks at Auburn could seemingly perform miracles, but even they could not have decided to build a totally new car on July 8 and send it off down the road a mere three weeks later!

The license lamp holder on the prototypes differed from the production cars.

CALIFORNIA, HERE WE COME!

Errett Cord, I believe, had approved the construction of the prototypes of the Cords-to-be in March 1935. Ames might have sent him Buehrig's drawings, or photographs of the red clay model. But Cord had never seen a full-size real car, and it's unlikely that he would have let the project proceed without such a viewing, and a personal test drive.

Since the prototypes had not yet had any serious testing, a shakedown trip was in order. And since E.L. was living on his estate in Beverly Hills, that became the destination of the first Cord to brave the highways.

Snow assigned George Kublin to head the three-man team making the trip. Joining him were two other Auburn men. They may have been Development Engineer Stanley R. Thomas and Experimental Garage Mechanic George P. Ritts. The car was E306 #2, a Westchester sedan. An Auburn memo describes this as "body #1." Snow's notes in the margin say "Car #2." This was indeed the

E306 #2 on its trip to Los Angeles. George Kublin is on the right. (Maynard Snow)

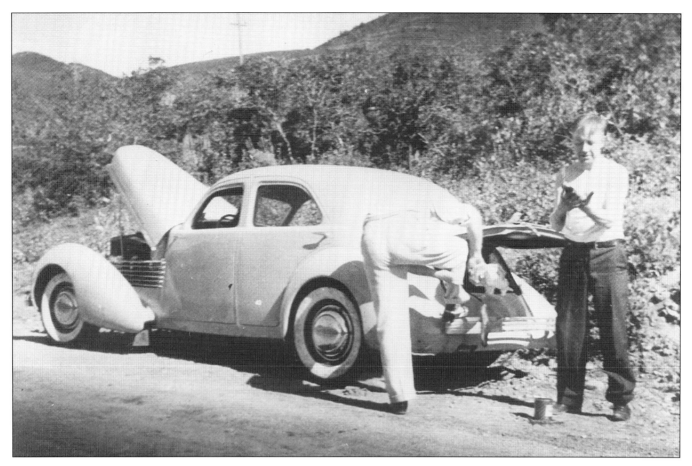

second front drive car built. It was, however, the first car with a Cord body. "Car #1" had an Auburn body.

The new Cord would certainly be an object of wonderment and amazement to those who saw it on the road or at rest. Auburn's marketing people preferred, though, that it not yet be associated with the company or with the name of the earlier Cord. (There was no Cord emblem on the transmission cover of this car, nor even a space for one.) The trip was scheduled for a weekend, when there might be fewer cars on the road. In the further interests of secrecy, the new car glided out of its Auburn garage at 2:30 in the morning on Saturday, July 27, 1935. It had 361 miles on the odometer.

The route they followed west from Auburn was essentially the old Lincoln Highway. By 1935 most of it had been designated US Route 30. The first highway experience in any untested machine is likely to be an interesting one, and so it was with E306 #2. Brakes proved inadequate, the transmission jumped out of second gear constantly, the electric shift failed periodically, and the Bendix universal joints rattled and knocked. Vapor lock was a cause for concern. The high compression ratio was causing detonation or knocking. They switched to Ethyl (leaded) gasoline, but the problem was still not completely solved.

But once rolling, the prototype car ran smoothly and fast, and halfway into the third day on the road Kublin was able to send this telegram:

Top: The back of the original snap-shot says "East of Salt Lake City." Looks like they'd been doing some repairs.

Bottom: We don't know where this photo and the photo on page 76 were taken, and whether it was on the way to or from Los Angeles. (Maynard Snow)

Left and middle: These views were taken on the Cordhaven driveways.

Right: Another Cordhaven shot, Harold Ames in the driver's seat.

```
N EVANSTON WYO JUL 29 1220P
R F FAULKNER
AUBURN AUTO CO.

HELLO ROY CHUCK FULL OF ENTHUSIASM AND HERE GOES A WEE BIT OF
IT. HAVE MADE MANY TRIPS WITH NEW CARS BUT WITHOUT
EXAGGERATION THIS IS THE GREATEST OF ALL RIDING QUALITIES
STEERING ROADABILITY TRANS SHIFT UNSURPASSED AND FOR EYE
APPEAL SEX APPEAL PUBLIC ACCEPTANCE ETC TREMENDOUS AND
STUPENDOUS IS EXPRESSING IT MILDLY. EVERY EFFORT TO AVOID
PUBLIC EYE IS FUTILE. THEY TRAIL US UP SIDE STREETS, COUNTRY
WAYSIDE FILLING STATIONS AND LITERALLY STAMPEDE THE CAR. THEY
GAZE IN WONDERMENT AT THIS SLEEK LOW CREATION. WE TELLEM ITS
A SPECIAL, A FOREIGN MAKE, A BULGAT, A NAZI, A WHOSAT ETC FOR A
PROMINENT HOLLYWOOD STAR ETC AND LOTS OF OTHER ANSWERS BUT
IN SPITE OF IT DOZENS OF PEOPLE ALL WALKS OF LIFE, DOCTORS,
FARMERS, MOUNTAINEERS, BUSINESS MEN ASSOCIATE THE CAR WITH
CORD AND SOME WITH AUBURN. HOW THE LADIES ADMIRE AND RAVE
OVER THE APPEARANCE, LOWNESS, GORGEOUS TRIM, BEAUTIFUL
INSTRUMENT PANEL, STEERING WHEEL AND OTHER GENERAL COM-
MENTS AND REACTION HIGHLY FAVORABLE. ALL INDICATIONS POINT
TO TREMENDOUS PUBLIC ACCEPTANCE. LOS ANGELES TOMORROW
AFTERNOON. HOPE YOU FEEL BETTER. REGARDS

                                                    GEORGE
```

Kublin didn't dwell on the bugs found to that point. As engineers, he and his colleagues recognized that most of the problems were superficial, and could be corrected in the production models. As Kublin promised, car and crew arrived in Los Angeles at 3:00 p.m. on Tuesday, July 30. Harold Ames, now fully behind the Buehrig-bodied project, had flown in from Chicago and was waiting for them. Perhaps he wanted to see the expression on Errett Cord's face.

The Wagner brake station in Los Angeles rebuilt the brakes, replacing drums, shoes and hoses. The local Auburn dealership pulled the cylinder heads and dropped the oil pan, so internal parts could be examined. One outer universal joint was replaced. Poorly operating Defiance brand horns were replaced with Delcos. To reduce the vapor lock, Harold Ames suggested installing an electric fuel pump. This was done, but the electric pump failed on the trip back.

E.L. Cord was not content to just lay eyes on his new namesake. He gave the car a vigorous workout on Los Angeles roads. He was

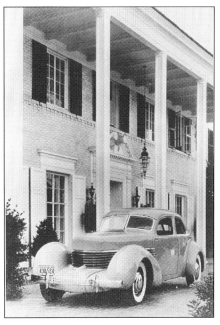

probably pleased with the car, but the engineers only recorded his criticisms. After all, that's what the test run was for. E.L. complained about body lean and tire squeal on hard cornering. The engineers worked up a rear roll bar and installed it. E.L. wanted to see the car stand a bit lower too; the engineers added 3/4 inch lowering blocks. Cord found the steering heavy on turns, and the caster return of the steering wheel too strong. He was also unhappy with fan noise. The engineers installed a fan with smaller blades.

Top: "Los Angeles," says the back of this snapshot. (Maynard Snow)

Bottom left: At the front door of Cordhaven. (Auburn-Cord-Duesenberg Museum)

Bottom right: Replacing the seized transmission in Dixon, Illinois. (Maynard Snow)

This photo of E306 #2 at Cordhaven was used in many of the trade reviews of the new Cord Front Drive.

Shifting became erratic after several days of local driving in Los Angeles. E.L. strongly recommended the use of a much larger battery, since reliable shifting was vital to the car's reputation. (Actually the problem was probably not battery size; it was the car's anemic third-brush type generator.) The original battery was replaced with a larger one, to make shifting more reliable on the trip home.

Photographs of the new Cord would be needed for catalogs and advertising. So during the six days that E306 #2 spent in Los Angeles, it was photographed in several settings at Cordhaven. These photographs were used in original and retouched form, until photos of the showcars and of production cars were available.

E306 #2 headed back to Auburn at 6:00 p.m. on Monday, August 5. Brakes still ran hot, and universal joints continued to knock. The lowering blocks caused the rear end of the car to bottom on hard bumps. The smaller fan, while quieter, made the engine run warmer. Ames' inboard headlights were a problem every night. The actuating cables broke, leaving the engineers to block the lids open at night. But the worst was yet to come.

Auburn's engineers were still months away from the discovery that an oil pump and pressure lubrication were needed to assure the transmission's long-term survival. The experimental drive unit used in E306 #2 had no such pump. After 4,900 miles on the road, on August 9 in Dixon, Illinois, a thrust washer broke, jamming the gears. Field remedies were not going to work, so a call went to Auburn for a replacement unit. Only five transmissions had been built by this date. Transmission #3 was in this California car. A driver brought transmission #5 out, and the engineers installed it. At midnight on Friday, August 11, E306 #2 came home.

The complete text of the reports written by Auburn engineers during and after this trip appear in Appendix ii.

9

COUNTDOWN

The peddler freight waited patiently on the long siding in Connersville, Indiana. It had been scheduled to pick up a string of loaded fifty-foot automobile cars in early evening. The freight cars still stood empty; apparently the load wasn't ready on time. Normally, the engineer would have continued on his route, picking up and dropping off cars at Connersville's many industrial sidings. But he'd gotten special orders as he passed the last station. They told him to wait for this load as long as it might take.

As the long October night wore on, the engineer and fireman discussed among themselves what could be so important as to influence the Baltimore and Ohio to change the train's schedule? And who could have pulled the strings needed to make this happen?

The news that the E306 project had been approved for production had been greeted with enthusiasm in Auburn. The town of Connersville, where the Cord Corporation's Central Manufactur-

The Auburn Automobile Company's manufacturing facilities in Auburn. The J-shaped building in the left foreground is the administrative headquarters and main showroom.

81

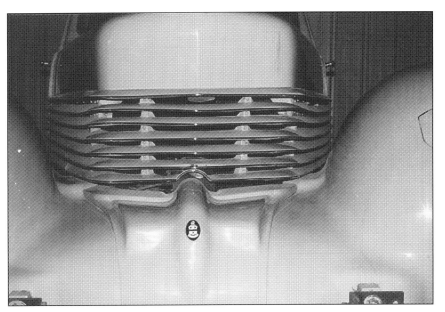

The grilles on the prototypes and show cars were bolted together. These braces are visible from the front.

ing Company subsidiary would likely assemble the new car, joined in the celebration. It wasn't long before reality set in. The Automobile Manufacturing Association permitted only production cars to be exhibited at the annual automobile shows; no prototypes were allowed. AMA required that at least one hundred cars be completed in order to meet the definition of "production." And there were only 116 days until the New York and Los Angeles shows both opened on November 2.

Although the wooden die models for the front drive sedan had been completed, there was no possibility, in that short period, of having all of the permanent tooling made, especially for body parts. If the new cars were to be exhibited at the auto shows, many of those parts would have to be made by hand, and assembled by hand.

Before we throw ourselves into the furious activity that was about to begin, let us pause to consider the magnitude of the task that the Auburn people had set for themselves.

There were about 10,000 parts in a car of that era. Many of these would be purchased as assemblies from outside vendors, others were stock fasteners and similar hardware. But the rest—they were not only unique to the new car but different from any other car on the market. Major assemblies, like the drive unit, had barely been tested yet. The unit body, later to acquire the French usage *monocoque,* was a new concept on these shores. Neither the unique suspension nor the complex universal joint arrangements had ever been on the road. Nor had the disappearing headlights or the four-speed version of the electric-vacuum gearshift. And while the Lycoming-built engine was the least of the problems, it too was an untested design. Tooling for only a few of the new car's parts had been prepared during the creation of the E306s.

Over the next sixteen weeks skilled workers proposed to hammer and cast and drill and machine and mill and turn thousands and thousands of parts, then clamp and weld and bolt and assemble them into the bodies and chassis' of one hundred new cars. They'd assemble and install one hundred engine and drive trains. They'd wire and upholster and paint one hundred bodies. And when all this was complete, they would deliver a selection of these cars to automobile shows on both coasts of the country. Then they'd take a deep breath, and pretend to the public that the exhibit was the result of a careful and thoroughly tested development process.

The original goal was to honestly meet the AMA's requirement for one hundred finished cars. Auburn placed orders with its parts suppliers for one hundred pieces or assemblies of mechanical and body components. Internal memos and letters to suppliers and to other Cord Corporation subsidiaries refer constantly to the need for speed to meet the AMA deadline.

> **Crank-and-cable**
>
> As soon as the retractable inboard headlights were fabricated and installed on E306 #2 it was evident to the stylists and engineers that this was not a workable design. The shakedown trip to California proved them right. So a new plan for a headlight mounted in the front of the fenders was quickly prepared. E306s numbers 4 through 6 used this design, as did the production cars.
>
> The crank-and-cable arrangement that operates the Cord's headlights has been described as a stop-gap plan because there wasn't time to work out an automated solution. Dale Cosper said later that the engineers actually worked on at least a dozen varieties of drives for the headlights. They tried electric motors and vacuum devices. The crank-and-cable mechanism was finally selected for production because it worked every time!
>
> The patent on the selected design is in the name of Herb Snow. Auburn made application for it in January 1936. It was granted in June 1938, ten months after Cord production ended.

The shakedown cruise of E306 mandated some exterior changes, which had to be incorporated in the showcars. Windshields were made larger, as were the rear windows. The problem of brake cooling received an engineering solution. George Ritts suggested that holes be punched in the chrome wheel covers to match the existing holes in the steel disc wheels. Buehrig was delighted with the improvement.

Airplane Development Corporation, another Cord Corporation subsidiary, manufactured the Vultee airplane. ADC used lead and zinc dies for limited production. These dies could be made quickly, and might be helpful in the creation of the showcars. Stan Menton was dispatched to the Vultee plant in California. His report in August suggested that such dies might be useful for the creation of limited run parts for the Auburn Speedster, but that other techniques would be more effective for the showcars.

The fragility of the project, and the responsibility everyone bore for its success, were highlighted by a note from Central general manager Arthur Landis to Harold Ames on August 20. Orders were being placed for tooling, and some minor cost adjustments were being discussed; some down, possibly a few up. Landis wrote, "We believe. . . it would be bad psychology to talk increased tool expenditure."

On Menton's recommendations, Herb Snow divided the work on the showcars' unit bodies between the Auburn plant in Auburn and Central Manufacturing in Connersville. Of the 100 first car bodies, 70 would be sedans, 15 convertible coupes, 15 phaetons. Sedan bodies would be assembled in Connersville. Phaeton and convertible coupe body shells, which would be assembled in Auburn, would be trucked to Central as they were completed. Final assembly of all the showcars would be done in Connersville.

Fabrication of body parts for the showcars was also distributed among both factories. For some parts, the creation of permanent tooling would take only a little longer than making the parts by hand. Making the permanent tools now would keep down the cost of the parts for the showcars, and provide a head start toward the beginning of actual production after the shows. These parts would be made in Connersville, since that's where the production cars would be built. So the showcars were assembled of a combination of handmade and production parts. Among the latter were the rear floor pan, instrument panel, rocker panels and glove box doors.

The major body panels for the showcars were fabricated in Auburn. Cowls for all the cars and the roof panels for the sedans were stamped

> **Scuff Guards**
>
> The chrome scuff guards at the lower front of the rear fenders were added after the return of E306 #2 from California, with badly pitted paint in this area.
>
> Most of the showcars had scuff guards cut square across the top. Some had a notch cut on the inboard edge. Auburn employee Ed Rudd remembered that when they tried to fit the hand-hammered guards to the show cars, some wouldn't fit properly. Cutting the notch made a better fit possible.
>
> The notch design was incorporated into the slightly taller production part.

> **Lloyd Henry Davidson**
>
> "Slim" Davidson was one of the hidden heroes in the Cord saga. He acquired the nickname which would be his for life when he started work at the Auburn Automobile Company in 1923. Nineteen at the time, he stood 6'1" and weighed 140 pounds.
>
> He started in the electrical components department, where wiring harnesses were made. He soon became its supervisor, then the head of plant maintenance. He was also responsible for the maintenance of E.L. Cord's home in Auburn. Davidson designed and built the welding fixtures for Auburn's first all-steel cars in 1934.
>
> He was in charge of the shop in Auburn that converted the early Auburn Speedster body shells to the 1935 and 1936 models. In that same capacity it was Davidson who supervised the construction of the E306 prototypes. Working out of Central Manufacturing Company, he would later manage the assembly of the showcars. He moved to Connersville in 1936; among his many projects there was the completion of four Duesenberg custom bodies by LeBaron, and their later fitting with Cord chassis.
>
> Davidson became chief experimental engineer for Central's kitchen appliance division in 1938. (His engineering education was acquired during his working years, primarily through correspondence schools.) After W.W.II, he was appointed Director of Product Development. A gifted inventor, he designed the dishwashers that were produced for many years on Connersville assembly lines, and sold under several brand names.
>
> Davidson retired in 1969. He died in Connersville in 1987.

Lloyd Davidson (Henry Blommel)

out on Auburn's Artz press. Parts for the front and rear fenders were hammered out in Auburn too. Fender halves were tack welded to hold them together, then shipped to Central in Connersville to be completed. Hood, louver assembly and transmission cover were made in Auburn, as were some smaller sheet metal pieces for the body and chassis. Panels and parts were trucked to Connersville, where all metal finishing and painting were done.

All wood parts were made in Connersville. So were the steel doors, floor pan, rear deck lid and door center pillar. All were fashioned by hand. From outside vendors came seat back assemblies, cowl ventilator, hood latch and props, and the headlamp mechanisms.

Columbia Axle of Cleveland, another division of the Cord Corporation, played an unheralded but vital role in the creation of the showcars. It was Columbia's task to assemble most of the mechanical parts for the show cars. This was done under contract to Auburn, and Columbia's role was specifically limited to the show car work. In addition to assembly work, Auburn ordered Columbia to check arriving parts against the specifications. In some cases, final drawings had not yet been done and Columbia was to approve these for initial production of the show car parts. Parts arrived at Columbia from all over the country. Transmission pieces came from a half-dozen different suppliers. Suspension and steering components came from six others. Where a supplier could not tool up in time, Columbia was instructed to fabricate the parts from scratch until tooling was ready. (Some components that would be made of cast iron or steel in the production cars were hurriedly cast in brass or bronze for the showcars.)

Auburn knew that there would be the devil to pay over the gearboxes. In a September 18 memo to Snow, S.L. Griffith issued a warning about Columbia's assembly of the transmission gears:

Detroit Gear will have representative at Columbia to assist in assembling these parts. We want to watch carefully and see that Detroit Gear's representatives assume all responsibility for this transmission gearset, as we want the operation of this transmission solely the responsibility of the Detroit Gear Co.

The eye of the hurricane was Central Manufacturing in Connersville. Here, Roy Weisheit's crew was literally putting together car bodies by hand. Body panels made here or shipped in from Auburn were checked against wooden bucks to be sure that they were shaped correctly. Still, there was enough variation from panel to panel, that each one fit only the car it had been made for. Porter Gorton developed a quick-and-dirty system for coding the parts so they eventually came together on the correct car. After one 70 hour week he went home to get some sleep, but was rousted out of bed by workers who said that they couldn't figure out which part went with which car.

In Auburn and in Connersville, workers rushed into the factory before dawn, and staggered wearily home after dark. 60- and 70-

This is probably E306 #5. These photos were taken in the factory's inner courtyard. The background didn't matter because it was to be airbrushed out. See the results on page 96.

Hand-painted

In 1935, Auburn employees were still rubbing out lacquer paints by hand! A memo from Arthur Landis to Harold Ames in August 1935 complained that Auburn was the only automobile manufacturer in the United States that did not own machine polishers. He pointed out that "this thing should take at least $2.00 off the paint cost of the automobile. In addition to giving a better and more uniform luster." There is no record of whether Ames approved the purchase.

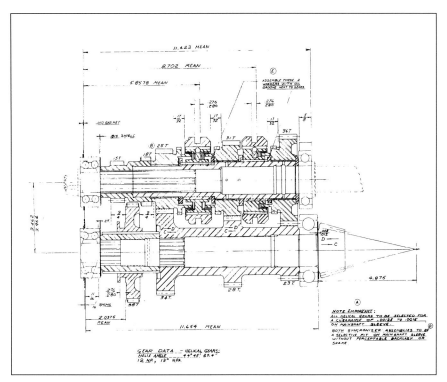

Transmission gear layout. (Thomas S. Pendergast)

hour weeks were the norm; payroll records show some men working as many as 90. In the economic night of the Great Depression, working people all over the nation were delighted with whatever work they could find. In these small cities in Indiana, for several months at least, happy days were here again.

Hurry, hurry, hurry! Memos from the main office exhorted Columbia, Central and Lycoming to greater and greater haste. The looming deadline of the November shows hung like a dark cloud over Auburn and Connersville.

In preparation for eventual production, some of the changes mandated by E306 #2's California shakedown were tried out on this car and on others of the five prototypes. A 1" taller radiator was installed on #2. So were Rzeppa universal joints. Windshields were made larger on #3. #5 was used for center of gravity studies.

All of the five E306 prototypes were rushed onto Indiana roads, driven hard to stress their mechanical parts and so discover weaknesses. They soon revealed themselves. Transmission #3, which was in car #2, seized on August 9 during its return trip from California. Its gears were returned to Detroit Gear and Machine on August 12, and retrieved on August 21. The reassembled gearbox was installed in car #1 on August 21, and stripped third gear on August 28. Transmission #2, in car #1, kept jumping out of second. It was replaced by transmission #1. #2's gears were sent back to DG&M on August 3, then reinstalled in car #1 on August 15 where they promptly stripped. Teeth broke on transmission #1 while it was in car #1. DG&M made new gears, and the gearbox was installed in car #4. It stripped third gear on September 9. It was a slapstick comedy of problems, but no one was laughing.

The small collection of transmissions and gears caromed from Auburn to Detroit Gear and Machine to Columbia Axle and back. Synchronizers jumped out of engagement, gears stripped under foot-to-the-floor acceleration, parts broke.

The five existing transmissions were no more than experimental. Different combinations of materials and bearings were used, in an attempt to determine in weeks what really needed months of

testing. One transmission carried its shafts in Timken tapered roller bearings. Another's gears spun on needle bearings. Final design decisions had not yet been made, and the countdown clock continued to tick.

On August 27, an ominous memo from Snow called for "15 sedans, 5 cabs [sic] and 5 phaetons to be completed by October 15." That was less than two weeks before the first cars would have to be shipped to the shows, and seems to indicate resignation to the inevitable—100 cars could not possibly be completed by show date.

Worse yet, the transmission problems had still not been resolved. Auburn's tests showed that friction losses in the new unit were extraordinarily high—at 3500 rpm, nearly triple that of the conventional gearbox in the Auburn 851. 40-grade engine oil was used as a lubricant during testing. One transmission got so hot that the oil burned to carbon, locking the shift mechanism.

The cabriolet with rumbleseat. (Henry Blommel)

As with many new and complicated mechanisms, there were flaws in the Cord transmission. Auburn simply did not have the time to correct them. Testing of the five experimental transmission/differential units went on until late October. Columbia had built no more, pending final design decisions.

So the new cars would go to the shows without transmissions. Car buyers were even more curious then about the mechanical aspects of a new car than they are today. When you're showing a car whose most advertised mechanical feature is front wheel drive, you can bet that little boys and grown men will be down on the carpet peering underneath at the unusual axles and drive mechanism. Some not-too-obvious means had to be provided for

E Pluribus Unum

To appreciate the logistics of the creation of the showcars, witness the fabrication of the front suspension arm. This simple part was assembled from components supplied by four different companies: Truscon Steel Company made the stampings; Pittsburgh Forging Company made the yoke; Powell Pressed Steel Company made the shock absorber bracket. The pivot tube came from a local mill. Columbia Axle riveted and welded the parts together.

The complex drive unit assembly was even more of a nightmare. For the 100 showcar transmissions only, Columbia cast the case. The patterns used were afterward sent to Spencer Heater in Williamsport, who cast the cases for the production jobs. Columbia cast the shift housing from temporary patterns. Parts arrived in Cleveland from all over the midwest. Gears came from Detroit Gear and Machine, shift rails and forks from Buckeye Forging Company, the stamped cover from Akron-Selle Corporation. Small bits and pieces were delivered from Auburn's own machine shops. Columbia made the differentials, and assembled the complete units.

The apron covering the missing front drive parts on this show car has visibly sagged a bit. (J.K. Howell)

concealing from view the cavity where the transmission was supposed to be. The device used was a fabric apron slung under the front end of the car.

Gordon Buehrig remembered that someone in management decided that the Cord exhibit at the New York show would be enhanced by the display of the quarter-scale red model. Since the

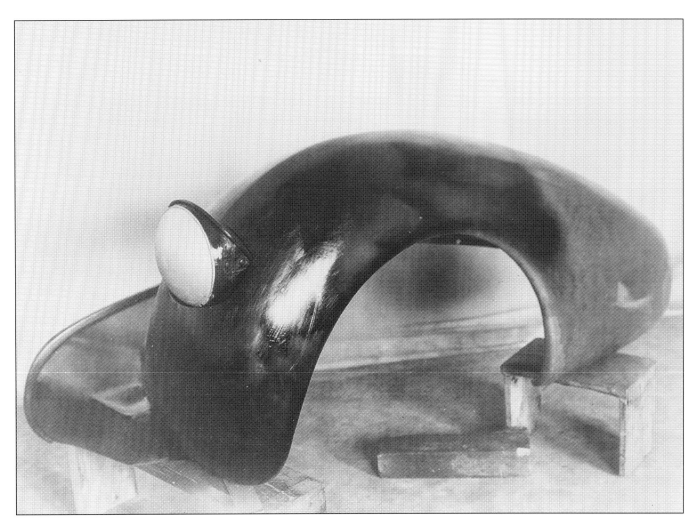

A fender mockup used for development of the production headlights. (Auburn-Cord-Duesenberg Museum)

"Gentlemen's Speedster" model was to be displayed as well, it's a good bet that the originator of the idea was Harold Ames. Vince Gardner and Dale Cosper designed and built cases for the models. The front, top and one end of each case was glass. On the back, bottom and other end were drawn the side, top and end views of the car. In other words, as Buehrig later said, the cars sat on their own body drafts, and illustrated how the design work had been done.

Trucks carrying parts and bodies continued to roar up and down US Route 27 and the two-lane state roads linking Connersville and Auburn. The frantic activity at Central Manufacturing reached a crescendo. The closing days of October were approaching, and still not nearly enough cars had been assembled to satisfy the AMA requirements.

Lloyd "Slim" Davidson was supervisor of Central's Experimental Garage, and overseer of the showcar project. At two o'clock in the early morning of October 29, Davidson looked out over the collection of nearly-completed cars. His educated eye told him that eleven of them were close enough to completion to be exhibited to the public. He selected those that would make up the Cord exhibit at the Los Angeles show. The shows in New York and Los Angeles both started on the same date, November 2. Since train time to the west coast was two days longer than to the east, the Los Angeles cars had to start on their way first. The way freight still waited on the siding in back of the Connersville plant, with the string of empty

The Final Drive Units

The experience of later owners showed that the transmission was a weak component of the production Cord. In the 1950s, the Auburn-Cord-Duesenberg Club set up a gear committee to determine what design flaws existed in the Cord drive unit, and what the solutions might be. The engineers on this committee found the design lacking in several areas. The helix angle chosen made the gears quiet, but weaker than they might have been. The high angle also caused excessive end thrust. This created excessive loads on the thrust washers, and contributed to the chronic slipping out of gear. The distance between front and rear bearings was too great, requiring shafts that were too long. This permitted the shafts to bend under load. And the gears themselves were narrower than they should have been, and so less able to adequately handle the torque of the Cord engine. Engineer Weaver, the committee suggested, might have been a conventionally competent professional. But on the Cord project he may have been a bit out of his depth.

Perhaps Mr. Weaver is being unfairly scapegoated. The transmission that he originally designed for E-278 had three speeds forward, with the top two synchronized. No drawings of this unit survive. Auburn's analyses in January 1935 showed that the production Auburn Model 851 could comfortably out-accelerate the new front drive car. When the Auburn shifted to the high ratio of its two-speed rear end, it left the new car in the dust on top speed too.

So, between February and May 1935 the decision was made to equip the new front drive car with a four speed gearbox, with fourth gear being an "overdrive" ratio. There was certainly no time to design a new transmission from scratch. So Weaver may have had to lengthen the case a bit, and narrow each gear a bit, and so fit in four sets of forward gears where there had been three before. Result: a too-long shaft, and too-narrow gears.

Conventional 3-speed transmissions were the norm in American cars. All of these had a "direct drive" in third gear. In the Cord transmission power passed through a set of gears no matter which speed was selected. And that, in turn, meant that there were a lot of gears churning around in the lubricant all the time. To compound the problem, the driven shaft in a gearbox of this design turns much faster than in a conventional transmission. In an overdrive gear like the Cord's afterthought fourth gear, the speed disparity becomes considerably greater. Here's an example. At an engine speed of 2,740 rpm, the equivalent of 80 mph in fourth gear, the driven shaft in the Cord transmission and the gears attached to it are turning at 4,288 rpm. These are the gears that run partially immersed in lubricant. In the conventional Auburn transmission in high gear at the same 2,740 engine rpm, the lower shaft, the one immersed in gear oil, turns at a comparatively leisurely 1,627 rpm.

The resulting friction and heat plagued the Cord gearboxes. The unique case design required by the combination transmission-differential made the friction and heat problems worse. Much more lubricant was required in this unit than in an ordinary transmission because it took a gallon of gear oil just to fill the well which housed the differential. Auburn's experiments began with seven quarts of lubricant. As it became evident that all that rapid churning by both transmission gears and the differential ring gear were responsible for much of the friction and heat, lubricant level was dropped. Good results were found with five quarts of oil, which put the equivalent of only one quart in the transmission. But this arrangement caused other problems. A memo on August 30 said that "When the transmission failed in road tests, it became evident that the gear bearings, and especially the washers between the different gears on the main shaft which transmit the thrust are not lubricated sufficiently." Maybe a pump was needed to provide lubricant under pressure to all the gear bushings and the thrust washers? Should the gears rotate on needle bearings, or would plain bronze bushings work better?

A memo to Columbia Axle on September 7 deals with continuing design changes for the drive unit. At that date the gears were still intended to run on needle bearings. It was not until October 26 that Auburn engineer Kaufmann could report that the best combination to keep friction and heat to acceptable levels was bronze bushings and pressure lubrication.

The auto shows in New York and Los Angeles were only seven days away.

Finding the E306s

The five E306 prototypes were numbered 2 through 6. There are photographs of some of them, but none identifies the subject by number. So figuring out which is which is a puzzle composed of equal parts of logic, guesswork and intuition.

Here's what we know:
1. All the cars had horizontally grooved bumpers; where used, bumper guards were similar in design to the 1935 Auburn.
2. #2 had a lower windshield and rear window than the others.
3. #4 and #6 were painted dark colors (black and clay rust, respectively).
4. #2 had no chrome gravel shield on the rear fender; all the others did. (Its shape was different from the showcars and from the production cars.)
5. Only the earliest cars had inboard headlights; probably #2 and #3.
6. The gas hatch on #2 was hinged at the bottom; the others, at the left side like the production cars.

Based on these suppositions, I've suggested an identification for each photo of an E306 published in this book.
   Hope you agree.

> ### Why "810"?
>
> The first record of the new car being referred to as the "Model 810" is in an engineering chart dated August 10. (Yes, that date can also be written as 8/10, but I think that's coincidental. Auburn usually began its model numbers with the number of cylinders, hence the "8." For the last two digits, they no doubt plucked the "10" from the same place as the "51," "52" and "53" on their Auburn models.) On August 27, a memo describing where the various bodies would be built refers to the "Model 810—Front Wheel Drive." Finally the next day an official memo from J.A. McBride announced, "The front wheel drive (Experimental models E294 and E306) will be known as Model 810."
>
> With a single exception, the term "Model 810'" was used only in internal memos and service bulletins. The exception was the Owner's Manual, which used "Model 810" on the cover. Ads and dealer brochures always referred to the car as the new Cord Front Drive, or some variation.

double-door automobile cars. A short time later, six Cords were pushed out of the 18th street door of the Central plant, then up the hill to the rail siding. It was two more hours before they were loaded into automobile cars. These special long freight cars could each accommodate four automobiles in the "half-deck" position; that is, tilted up at a 45 degree angle by a loading mechanism in the rail car. Engines were not filled with oil for this kind of loading, and these Cords had no transmissions from which lubricant might have leaked. The local freight pulled away, hauling its load to the interchange point where they would be coupled into a fast freight headed for Los Angeles. The cars for the New York show would leave two days later.

The display cases for the quarter-scale models were not ready until after midnight of October 31. Gordon Buehrig had to go to the show anyway, so he was assigned the task of transporting the exhibits. Auburn was doing some business in the undertaker market, and had a hearse-like utility vehicle they wanted to show to dealers in New York. Buehrig used one of these to drive the cases and models to New York. Perhaps no one noticed the omen.

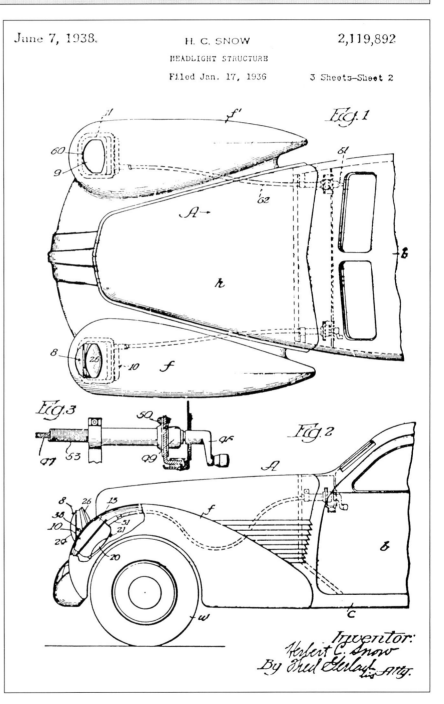

Herb Snow's headlight patent. (US Patent Office)

The Push-button Shift

The Stromberg EE-15 carburetor used on unsupercharged Cord engines was fitted with an automatic choke of a new design. Stromberg's facilities in South Bend, Indiana, included a "cold room." This was a space large enough for a car to be driven into. Temperatures could be dropped to zero or below, so cold-weather performance could be evaluated. In October 1935, Stromberg asked Lycoming to bring one of the new Cords to South Bend so the new automatic choke could be tested on a car.

Prototype E306 #3 was then in Williamsport. A crew including "Rick" Rickenbach, another Lycoming engineer and an Auburn engineer set out for South Bend. The Auburn man drove.

This prototype was equipped with the stock Bendix gearshift mechanism, but it was actuated by push-buttons located at the center of the steering wheel. All other makes of cars had their horn button in that position. At one point another car cut off the Cord. The driver, by habit, put his hand on the steering wheel center to blow the horn. Instead, he hit several shift buttons simultaneously, and the gearshift went into neutral. Rick remembers that it took quite some time for the Auburn engineer to sort out the wiring and switches and get the car moving again.

The push-button gearshift wasn't seen again.

Artist's conception of the push-button gearshift, based on August Rickenbach's recollection. This experimental device was fitted to E306 in September 1935. (Ken Eberts)

10
SHOWTIME

Over 58,000 eager viewers crowded the Los Angeles auto show on the weekend that it opened. The New York show was similarly mobbed. The response of the public, the dealers and the press to the new Cord was everything its Auburn parents might have wanted.

The shows were the big news in November 1935, and every newspaper and magazine carried page after page of stories about them. A surprising number of these were illustrated with a photo of the new Cord, photographed at the show or supplied in the P.P. Willis agency's press kits. Nearly every article described the Cord. All this priceless press coverage went to a car whose annual output resembled one day's production by General Motors. Why?

To understand the impact of the Cord's now-familiar lines, you must examine them in the light of the norms of the day. The Cord's

Artist's conception of the red scale model in its case at the 1935 New York Automobile Show. (Ken Eberts)

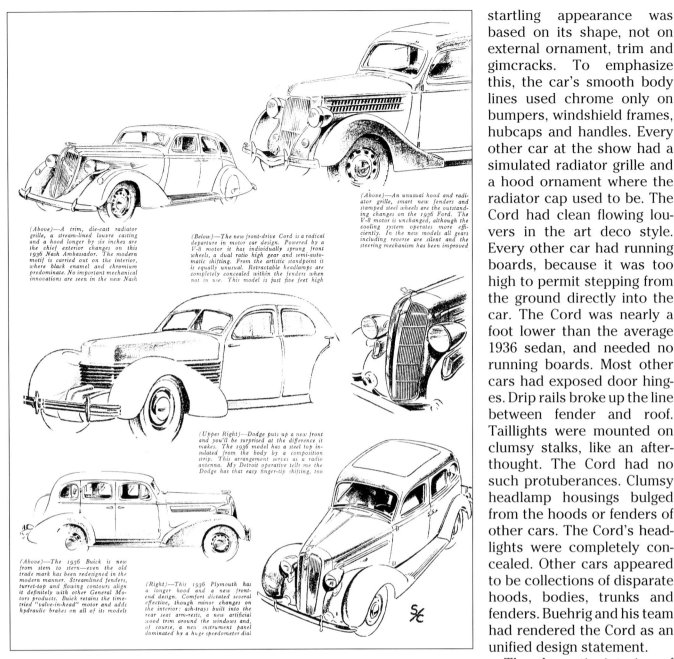

(Above)—A trim, die-cast radiator grille, a stream-lined louvre casting and a hood longer by six inches are the chief exterior changes on this 1936 Nash Ambassador. The modern motif is carried out on the interior, where black enamel and chromium predominate. No important mechanical innovations are seen in the new Nash

(Above)—An unusual hood and radiator grille, smart new fenders and stamped steel wheels are the outstanding changes on the 1936 Ford. The V-8 motor is unchanged, although the cooling system operates more efficiently. In the new models all gears including reverse are silent and the steering mechanism has been improved

(Below)—The new front-drive Cord is a radical departure in motor car design. Powered by a V-8 motor it has individually sprung front wheels, a dual ratio high gear and semi-automatic shifting. From the artistic standpoint it is equally unusual. Retractable headlamps are completely concealed within the fenders when not in use. This model is just five feet high

(Upper Right)—Dodge puts up a new front and you'll be surprised at the difference it makes. The 1936 model has a steel top insulated from the body by a composition strip. This arrangement serves as a radio antenna. My Detroit operative tells me the Dodge has that easy finger-tip shifting, too

(Above)—The 1936 Buick is new from stem to stern—even the old trade mark has been redesigned in the modern manner. Streamlined fenders, turret-top and flowing contours align it definitely with other General Motors products. Buick retains the time-tried "valve-in-head" motor and adds hydraulic brakes on all of its models

(Right)—This 1936 Plymouth has a longer hood and a new front-end design. Comfort dictated several effective, though minor changes on the interior: ash-trays built into the rear seat arm-rests, a new artificial wood trim around the windows and, of course, a new instrument panel dominated by a huge speedometer dial

The Cord stood out from its contemporaries. (Ron Irwin)

startling appearance was based on its shape, not on external ornament, trim and gimcracks. To emphasize this, the car's smooth body lines used chrome only on bumpers, windshield frames, hubcaps and handles. Every other car at the show had a simulated radiator grille and a hood ornament where the radiator cap used to be. The Cord had clean flowing louvers in the art deco style. Every other car had running boards, because it was too high to permit stepping from the ground directly into the car. The Cord was nearly a foot lower than the average 1936 sedan, and needed no running boards. Most other cars had exposed door hinges. Drip rails broke up the line between fender and roof. Taillights were mounted on clumsy stalks, like an afterthought. The Cord had no such protuberances. Clumsy headlamp housings bulged from the hoods or fenders of other cars. The Cord's headlights were completely concealed. Other cars appeared to be collections of disparate hoods, bodies, trunks and fenders. Buehrig and his team had rendered the Cord as an unified design statement.

The dramatic interior of the Cord bore out the promise of the exterior. The large round airplane-style instruments and the switches styled like airplane throttles all spoke to the most modern of 1936 transportation icons. The engine-turned dash panel was a true expression of the art of the Machine Age. The lush broadcloth interiors in colors contrasting with the exterior paints were even more impressive when compared to the beige and grey mohair of other cars.

Most cars were displayed in the dark blues, maroons and black that characterized passenger cars of the mid-thirties. The New York show Cords dazzled in sapphire blue and sienna brown. (Production cars were not offered in these exotic colors.)

The madcap circus was not over yet. The cars that Lloyd Davidson had selected for shipment to the the shows were the closest to ready, but still not fully completed. Roy Weisheit

> **Whodunnit?**
>
> The sensation caused at the New York show by the local product did not go unnoticed in Indiana. In November 13, a syndicated article on the auto shows appeared in a Fort Wayne newspaper. It gave credit for the Cord styling to independent stylist Count Alexis de Sakhnoffsky. Buehrig told Faulkner he would resign unless the error was corrected. A small press release was sent out stating that credit should go to an Auburn employee named Gordon Buehrig. This was the first public credit Buehrig had ever received for his work at Duesenberg or Auburn.

remembered that he spent much of November in the air in the company's Stinson Tri-Motor, carrying parts to the shows. They were installed on the showcars on the floor, after each show closed for the day.

None of this frenzy was evident to the showgoer, who did not need to dissect the Cord's design to find out why it was right. He just knew it as soon as he laid eyes on it. And he knew he had never seen anything like it before. So he crowded around the small Auburn exhibit. It's reported that showgoers stood on the bumpers and running boards of competing cars to get a better look at the Cord, so dense were the crowds. The reviewer from *The Los Angeles Times* found "a huge circle of absorbed onlookers" at the Cord exhibit.

Color brochures to be distributed at the shows had been prepared by the P.P. Willis agency. The seven detail photographs were all of E306 #2. Art depicting the three models—The Sedan, The Convertible Phaeton Sedan and The Convertible Coupe—was probably by Herb Newport. Two styles were shown in red, a color not available in the production cars. The car was referred to only as The New Cord. The brochures vanished quickly. Requests for more were being received by dealers and by the factory for weeks after the shows.

The press truly fell all over itself in making the new Cord the star of the shows. *The New York Times,* in its special feature on the show that had opened the previous day, devoted 16 inches to a description of all the new Chevrolet models, and 14 inches to the Dodge line. Cord got 12. A three-column publicity photo of E-306 #2 headed one page. *The Los Angeles Times* would not be outdone. "Cord Front Drive Exploits New Principles In Automobile Construction," screamed its headline, "Real 'Pleasure Car' Attracts Interests of Crowds at Auto Show." The new Cord, said the article, ". . . will unquestionably set a trend that will affect motorcar design for many years." *The Times* also headed another article with a photo of a Cord, taken at the show.

Top: The cover of the catalog distributed at the 1935 shows.

Bottom: A beaming E.L. Cord at the Cord exhibit at the New York show, with Fred Zeder, chief engineer of the Chrysler Corporation. (Ron Irwin)

A striking effect is obtained by eliminating the usual radiator grille and extending the louvers around the front

Above—Rear lamps and gasoline filler are flush with the rear deck while the license plate and its lamp are mounted flat against the luggage door at the center. Below—The engine is accessible by raising the hinged hood. The louver assembly is readily removed

Disappearing headlamps are optional at extra cost. They are shown in raised position at left and flush with the fenders below

New Cord Front Drive

Lively Fast Car of Novel Construction Has 125 hp Vee Eight Engine and Weighs Only 3500 Pounds

ONE of the most interesting automobiles seen in many years is the new Cord front-drive built by the Auburn Automobile Co. Not only does it present new thoughts in styling but many of its features of construction are equally novel. It is equipped with a 125 hp vee eight engine, has a wheelbase of 125 inches and yet weighs only 3500 pounds because of its compact front-drive power plant and its unique all-steel body with in-built frame.

It has a four-speed transmission of new design in which second, third and fourth speeds are equally silent. Third corresponds to direct in other cars while fourth is comparable to overdrive since it slows the engine down about 30 per cent. Front wheels are independently sprung on longitudinal arms in connection with a single transverse leaf spring. Rear springs are half elliptics supported by a tubular rear axle. It is said that the total unsprung weight has been reduced 39 per cent.

The elimination of the conventional propeller shaft and banjo rear axle—thanks to front-drive—permits locating the floor only 10⅝ inches from the road and therefore, although headroom is the same as in the Auburns, the overall height of the closed cars is only 60 inches while the convertibles are 58 inches. The roof of the car comes to the average man's shoulder. The floor is so low that running boards are superfluous.

Headlamps fold into the fenders. The elimination of their air resistance is said to add five miles to the speed of the car bringing it close to 100 mph. Each lamp is hinged to the fender near its base and is swung up and down by a crank on the instrument panel which turns a flexible shaft within a flexible tube running to a worm which is meshed with a sector on the lamp.

This unusual looking automobile was designed on the principle that every detail had to have a practical reason for its adoption, although in each case the result has added to the striking appearance of the car as a whole. Many examples are given below. First is the use of louvers extending around the hood compartment. They permit a ready flow of air

From the November 1935 show issue of MOTOR *magazine*

The Cordhaven Album

Publicity photographs of all the 1936 body styles would be needed for the ad campaign and for sales literature. The only suitable photos available until November 1935 were the ones taken during E306 #2's hurried trip to Cordhaven. The estate was still the best place for photography, so it was there that the photographer and his crew, the models, and three transmission-less Cord showcars assembled after the Los Angeles show ended on November 9.

Fifteen of the photographs taken during this session appeared in ads and in sales material. A total of forty-four shots were printed by the photographer, and presented to E.L. Cord afterward. Many of the unused photographs were similar to those that appear in the ads, with a slightly different camera angle or model position.

Published on pages 99 through 101 for the first time is a selection from the photos that were not used.

Ultramodern in appearance, they said. The result of ten years of experimentation and development, they said. Roy Faulkner's explanation of why independent manufacturers could best offer innovation was reprinted in full in *The New York Times* and other papers.

The woman's viewpoint was expressed rather dramatically in *The Los Angeles Times:* "If you have a modern home you can just buy a new Cord and move it in the front room for an extra piece of furniture. For of all of the cars the new Cord is a thing of beauty in design, color, and mechanical perfection."

The Chicago Show began on November 16, 1935. Every make of car was under one roof at the International Amphitheatre at Halstead and 43rd Streets. This show was sponsored by the Chicago Automobile Trade Association, so Ford could display too.

Top: At the Chicago show. (Randy Ema)

The stand of R.S.M., Auburn's London distributor, at the Olympia Motor Show in Kensington. (Chas. Bowers & Sons)

All the car exhibits were on the second floor. The Auburn exhibit included the new Cord; Duesenberg exhibited separately. (Auburn showed its ambulance and a diesel-powered truck on the main floor.) The 3,400 square foot Auburn space was among the smaller ones. Only those of Reo, Pierce Arrow, Duesenberg and the exhibit shared by Willys and Hupmobile were smaller. It's likely that the Cords shown were completed during the two weeks since the Los Angeles and New York cars had left the factory, and came directly from Connersville.

The Auburn/Cord exhibit was small, stuck in a corner, hard to find. And mobbed. Chicago showgoers had read of the Cord's reception in New York and Los Angeles, and wanted to see for themselves.

The trade press, normally more discerning than the daily newspapers, was enthralled. Multi-page articles about the Cord, prepared directly from Willis' press kits, appeared in every automotive trade publication. Accompanied by the now-familar retouched photos of E306 #2, the articles repeated the advantages of front wheel drive, and all but suggested that mechanics start studying now to be prepared for the wave of the future.

Normally reserved British publications succumbed as well. *The Automobile Engineer* ran an 8-page article in its January, 1936, issue which purported to be an engineer's analysis of the Cord's engine,

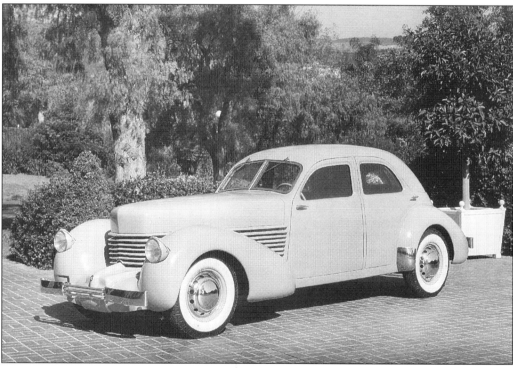

front drive and unit body. Author Herbert Chase, M.E., M.S.A.E., meticulously praised each and every detail of the Cord's engineering. Auburn distributed copies of the article to their dealers. And well they might. A copy of Chase's typewritten manuscript shows Herb Snow's penned corrections, additions and suggestions in the margins!

The publicity world had vouchsafed millions of dollars of free publicity to Auburn. The ball was now in the Indiana company's court.

Top: The Convertible Phaeton Sedan. The location is not Cordhaven, but a friend's home nearby. There wasn't time to fit the open show cars with windshield wipers. (Mrs. Errett Lobban [Virginia] Cord)

Bottom: The Westchester sedan with headlights open. (Mrs. Errett Lobban [Virginia]Cord)

The Convertible Coupe, in a rare rear view. This car had a T-handle to lift the top lid, matching the one for the rumbleseat. (Mrs. Errett Lobban [Virginia] Cord)

Top: The Convertible Phaeton Sedan at Cordhaven. Pete Willis, head of Auburn's ad agency, is in the driver's seat. Don Smith, Virginia Cord's brother-in-law, stands at the hood. (Mrs. Errett Lobban [Virginia] Cord)

Middle: The Convertible Coupe, in another unusual shot showing it carrying four people, two of them in the rumbleseat. (Mrs. Errett Lobban [Virginia] Cord)

Bottom: The press preview for the Chicago Auto Show took place nearly two weeks before the show itself. All of the completed Cords were then on the floors of the New York and Los Angeles shows. Auburn couldn't risk missing out on the publicity, so it sent prototype E306 #3 to be photographed and admired in the lobby of Chicago's Merchandise Mart. The show press release referred to the car's manufacturer as "The Cord Company of Auburn, Indiana," and made specific mention of the inboard headlights! When the Chicago Show opened on November 16, the showcars from New York were ready for display, and E306 #3 returned to obscurity. (Auburn-Cord-Duesenberg Museum)

The Bronze Cords

Cord dealers at the auto shows were promising delivery of the new cars by Christmas. After all, the factory had told them to.

The Auburn sales department understood that it would be fatal to sales if customers ready to put down a deposit on a new Cord were told to keep their money until a firm production date was set. So eager customers were told that delivery would begin within six to seven weeks after the show. This was a deliberate untruth. The factory knew it and the dealers knew it. The only people who didn't know were the customers.

We know that the factory had no intention of delivering cars by Christmas because in October, weeks before the auto shows opened, William Weiss—the president of a small firm called the Rotary Company of Buffalo, New York—visited Auburn at Harold Ames' invitation. He met with Gordon Buehrig, examined the quarter-scale red model, and left with drawings of the new Cord. (He was probably sworn to secrecy too.)

Rotary's entire workforce consisted of William Weiss and his brother. They had just been given a contract to produce 100 1/32 scale bronze models of the Cord sedan. In a brief two months, they had to develop patterns and produce bronze castings which had to be hand detailed, polished and assembled. Door lines, glass edges and windshield frame outlines were hand engraved into the castings. Louvers, bumpers and door handles were separate plated castings. So were the wheels. Each jewel-like result was mounted on a slab of onyx from California. The models were finished just in time to be shipped to Auburn by Railway Express, so dealers could get them to customers before Christmas with a note of apology from Auburn. It was the story of the creation of the full-size Cord repeated in 1/48 scale!

Since the models were ordered long before the shows took place, Auburn had no way of knowing how many would be needed. One hundred seemed like a nice round number. If Faulkner's figures are to be believed, many customers who had put down deposits received neither a big nor a little Cord for Christmas.

Rotary actually made 102 models. The extra two were mounted on a larger slab of marble with two penholders, as a desk set. One was sent to Harold Ames; William Weiss kept the other.

11

OUT THE DOOR

The showcar Cinderellas were off to the ball. Auburn could now give full attention to preparation for production.

Roy Faulkner wrote on December 9 that 7,639 requests had been received by mail for information on the new Cord. While a substantial number of these must be discounted as requests from literature-collectors, this was still a most impressive response. At a meeting with the press, Faulkner was optimistic about a resurgence in sales prospects for expensive cars. He offered the startling opinion that "it is no longer fashionable to pass as being poor"! Further, "Reception by the public of the new Cord has been so enthusiastic that the Auburn Automobile Company has doubled its projection for 1936. Production on the new Cord is being

Test driver Ab Jenkins with E306 #4, in front of the main showroom of the administration building in Auburn. (Henry Blommel)

103

Cords were coming off the line on February 15, 1936, during the training of Auburn-Cord service managers. 1936 Auburns were still being produced. Two are on the line at the left. The sixth and seventh cars in the Cord line are Auburns too! (Henry Blommel)

speeded up and the anticipated shipments for January will total around 1,000 cars." All that remained was to build them.

The pressure on Auburn and on Central Manufacturing Company was now even greater than it had been before the shows. Showcars only had to look good. (Their Connersville assemblers thought of them just as pretty shells; Roy Weisheit referred to them as "skins.") The untested production cars would have to stand up in the real world. And the involuntary testers—customers who had just paid a price that could have bought them two Oldsmobile touring sedans—were not likely to be forgiving.

There was no more time to spend on the troublesome transmission, and it received only minor improvements. A lubricant pump had been specified in September 1935. The first design used was an archimedes screw type. Pressure was inadequate, and for production this was replaced by a small gear pump. (Five hundred cars would be on the road when it was discovered that pumps were failing because of the volume of steel shavings from gears and synchronizers that were being deposited in the lubricant. The quick fix was a filter inserted into the oil line. Because the easiest way to do this was to put the filter after the pump, particles still reached and damaged the pumps.)

It was engine manufacturer Lycoming who built the Cord transmission. Lycoming had never put together a transmission before, and had to learn how under the most difficult of circumstances. Patterns for the drive unit cases were moved from

Columbia to Spencer Heater in Williamsport. The cases were cast at Spencer, as were the shift housings. (Bell housings, connecting the engine to the drive unit, were cast here too.) The raw castings were moved the several blocks to Lycoming's Building 8. There they were machined to the specifications provided by Auburn. Gears arrived from Detroit Gear and Machine, other parts from a variety of vendors. Lycoming workers assembled the gearboxes, following instructions provided by DG & M and Auburn. A clutch purchased from Long Manufacturing Company was bolted to the flywheel of each completed V-8. The bell housing was installed, and the drive unit attached. The engine-drive unit assemblies were now ready to ship to Connersville by rail.

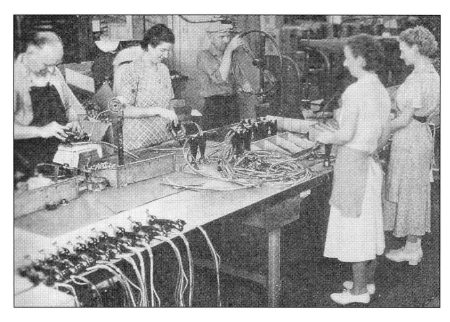

Bendix made some parts for the Cord in its 1,250,000 square foot plant in South Bend, Indiana. Assembled in this room were the selector units for the "Finger Tip Gearshift." The ones on the bench are for the Hudson and Terraplane. (Ron Irwin)

Essentially, what was actually manufactured on the master body fixtures of Central Manufacturing Company in Connersville was a body shell, from steel stampings made on Central's presses. Some smaller parts were fabricated there too. The body was painted on Central's very modern paint line, and trimmed in Central's own upholstery shop. But nearly every mechanical part used in the Cord was received at Central by rail or by truck. So were castings and hardware, and the hundreds of fasteners used in each car. Central's work force built many of the sub-assemblies from these parts, then put them together into a finished automobile as it moved down the assembly line.

Actually, even the sheet metal stampings for the bodies of the earliest Cords were made elsewhere and shipped in to Connersville. Central was very busy; they were still producing Auburns, building the kitchen cabinets that paid the bills, and doing outside stamping work. So the job of creating the dies for the Cord body was farmed out to tool and die companies all over the midwest. Body dies were due to arrive in Connersville in December. Setting up the dies in the presses and making test stampings would have delayed the start of production of body parts for weeks, and the beginning of car production for even longer. Auburn didn't have that luxury.

So all the orders for tooling included a supplemental order for 200 parts or sets of parts, made from those dies. The panels for the front fenders were stamped out by Koestlin Tool and Die in Detroit. Rear fenders came from Mullins Manufacturing Company in Salem, Ohio, of trailer fame. Mullins also made most of the roof panels, the hood, and the cowl. Windshield and dash parts came from Paxson Tool and Die, also in Salem. Superior Tool and Stamping Company of Chicago made the trunk lid panels. Louvers were produced by F. Joseph Lamb Company of Detroit. Other parts came from Holsclaw Brothers, Powell Pressed Steel, Crescent Tool and Die, Gibbons Tool and Die and Practical Die and Stamping Company.

Bendix tested gearshift selector units before shipment. (Ron Irwin)

Stub frames were shipped in from Midland Steel. The engine/drive unit package arrived from Lycoming. Rear axles came from Truscon Steel, spindles from Atlas Drop Forge. Truscon, Pittsburgh Forging and Powell Pressed Steel provided the parts for the front suspension arms. Gearshift parts came from Bendix' Bragg-Kliesrath Division. Instruments were by Stewart-Warner, headlight switches by Soreng-Manegold. Delco provided the horns, RBM the relay for them. Kingpins came from Canton Forge. Headlight and taillight parts were from Corcoran-Brown. The radio was a Crosley model, as was the under-car antenna. The muffler and exhaust resonator came from Powell. The radiator was a product of Jamestown Manufacturing Company. Atlas drop forge made all the steering arms. Wheel bearings arrived from New Departure and Timken. Atwood Brass made the large packing nuts used on the universal joint housings. Most gaskets came from Detroit Gasket, most seals from Chicago Rawhide. Wheels were made to Auburn's specifications by Motor Wheel in Detroit. Hubcaps came from Lyon Manufacturing, tires from Goodyear and Firestone.

Universal joints and front axles were from Bendix and Gear Grinding, Lockheed-designed brakes from Wagner. Seat frames were provided by Ashtabula Bow Socket Company. Door handles, trunk latches, hubcaps and wiring harnesses arrived from cities all over the midwest. Central's crew turned all these parts into a finished automobile.

The first Cord 810 came off the production line in Connersville on January 31, 1936. It was test driven in Connersville, then driven to Auburn for evaluation. Let us pause to look at this car, paying attention to those details in which it differed from the showcars and the prototypes.

The production FB engine's aluminum cylinder heads were polished in the early examples. Exhaust manifolds were coated with black porcelain enamel. Concerns about induction cross-firing notwithstanding, spark plug wires were encased in smoothly curving plated looms. The intake manifold, thermostat housings and the oil filler were aluminum too. (Several parts originally specified as aluminum castings were quickly replaced by stamped steel. The timing chain case was one, the oil pan the other. The latter was changed for both practical and economic reasons. The steel pans cost less. They also dented, rather than cracked, when the low-riding Cord hit a high curb or a rock.)

Like other engine manufacturers, Lycoming purchased engine accessories from other makers. For the Cord V-8 Aluminum pistons were provided by Bohn. Piston rings were by Sealed Power Corporation; the water pump was made by G & W Manufacturing.

The earliest FBs were equipped with Stromberg EE-1 carburetors, similar to those used on Lycoming-built Auburn straight 8s. Production FBs carried a Stromberg EE-15. (Series EE carbs were also used on some models of Cadillac, Ford, Buick, Chrysler, Pierce-Arrow, Nash and Studebaker.) The model provided on the Cord used a vacuum-operated piston to control fast idle and choke opening after cold starts; other Stromberg designs used a stepped cam for this purpose. Further choke opening was governed by a heater wire built into the automatic choke housing. Auburn was concerned about some of its customers

This montage includes a portion of the photo on page 104. The workmen have been dressed up by photo retouching; their plain white shop coats now say "CORD." (Ron Irwin)

being uncomfortable with the new-fangled automatic choke; a manual choke lever was part of the instrument panel, and was later used by dealers to hook up an optional factory retrofit kit.

The crankcase ventilation system in the earliest FB engines admitted air through a filtered breather mounted on the top of the timing chain case. A design change was made to create a mild negative pressure in the crankcase, to assist in keeping oil inside where it belonged. Makeup air was now to come in through a controlled opening in the oil filler cap.

Generator, starter and distributor were Auto-Lite products. An experimental test curve shows slightly higher FB engine output using a Delco distributor. Cost was probably the deciding factor. The Auto-Lite distributor approved for purchase had centrifugal advance only, while many cars of that era were equipped with a supplementary vacuum advance. The 810's third-brush generator design was in its last years of production. It incorporated a so-called "two charge" regulator mounted on top of the generator. The name related to the fact that this device could provide only a high or low charging rate, based on engine speed and battery charge. Its obsolescent design actually increased the charging rate as the battery approached full charge. In an attempt to reduce battery overcharging, many highway drivers of the 1930s kept their

The Engineering Notes

Auburn engineers worked at improving their product from the day production of the Cord began to the day it ended. The myriad changes and improvements were typed on 8 1/2 X 5 1/2 blank paper and collected in two small three-ring ring binders.

In 1959, Auburn-Cord-Duesenberg Club member Joe Knapp located these binders at the Auburn-Cord-Duesenberg Company. They provide us with a window into the happenings at Auburn and Connersville during the eighteen months of Cord production.

No item appears to have been too small or too large to be logged in. The engineering notes give us a record of when changes and improvements were made during production. The constant, often frantic, experimentation is revealed by these notes. Many are staid, some are bizarre, a few are hilarious. Together, they present a picture of a small company battling the big guys, and trying to look dignified while tripping over its own shoelaces.

Here's a sampling of the hundreds of items in the log. My comments are in italics.

12/3/35 (Received from Lycoming) We have changed the finished specification on all plated parts from chromium plate to cadmium plate. This change was made to reduce cost, and was made at your request. *The first production car was still two months away. Apparently only the very first batch of engines delivered by Lycoming used chrome trim.*

1/30/36 *Some bumpers had been delivered in which a spacer rattled.* All bumpers in our stock should be hit hard enough to see if this washer is loose.

2/21/36 In rechecking our crankcase capacity. . . we find there should be eight (8) quarts of oil instead of seven (7) as previously announced. *The dipstick Lycoming used for the FB engine was a stock Auburn part, and showed "full" at seven quarts. It was later revised.*

3/10/36 Serial 810-1330 was the start of 1 1/4 pnts of Kendall oil and 1/4 pint Schaler oil in inner joint. . . . 810-1394 ended the use of Schaler oil in inner joints. . . . 810-1446 they started using Schaler oil in inner joints again.

4/8/36 . . . rear roll bars were eliminated in production. . . *These bars had been added based on E.L. Cord's dissatisfaction with E-306 #2's lean on curves. They turned out to cause squeaks and rattles.*

5/28/36 With reference to your complaint of horn button jumping off the steering wheel. . . *The problem was found to be the Tenite casting, but Auburn was stuck with them. Some changes were made in the part onto which the horn button screwed, with some success.*

6/3/36 We have incorporated in production a change to eliminate the clutch and brake pedals sticking in the guides. This [change] has proven very satisfactory in several experimental cars, one of which has gone over 1,500 miles. *For the Cord, this was extensive testing!*

6/25/36 First car with interlock transmission installed 100% on line. Cars still in the factory and at dealers were retrofitted.

7/15/36 In building up the four 133 1/2" wheelbase Front Drive Cords with the Duesenberg bodies, we will use the damaged frame which is here in Connersville stock and will dismantle the Auburn job built up by the Cummins engine company with the Diesel Motor. Mr. Pommert is now obtaining two damaged frames which Mr. Davidson and I found in the junk yard at Auburn. These will be shipped by him and will complete the necessary order for four frames. *Talk about recycling!*

7/15/36 *Two engines with a new type of piston were shipped to Connersville by Lycoming.* As we have no provision for testing these engines, it has been suggested that they be installed in Cord cars on the assembly line and a record of the serial numbers be kept and sent to Mr. White at Auburn, so he may keep a record of their action in the field. *The customer as tester again!*

3/11/37 First supercharged car with Stromberg Carburetor having latest choke.

7/2/37 So all cars could be run off the assembly line today, the following inner joint housings were delivered from salvage to the assembly line without repairing:

2 — used, removed from used car with no information as to reason.
4 — removed from experimental cars for leaks.
3 — damaged oil seals, leak.
1 — oil seal missing.

The serial numbers of the cars on which they were installed were given, with the request that they be held in the test department until new parts arrive.

8/17/36 All cars after this. . . had Temperature Gauge adjusted to show 6 to 10 degrees lower temperature of water. *Customers were upset when the gauge showed boiling even though the car wasn't. Radiator water temperature was indeed lower than the original gauge indicated.*

Albert H. McGinnis, Auburn's Vice President in Charge of Production, with a new 810. (Henry Blommel)

headlights on in the daytime. Both Delco-Remy and Auto-Lite had already brought to market the modern shunt design generator, with separately mounted regulator including a true vibrating voltage regulating element. Cost was probably the deciding factor again.

A touch of the latest in automation was provided by Startix. This black box replaced the starter solenoid, and automatically restarted the engine if it stalled. The unit was wired through the clutch switch, so starting only took place when the clutch was depressed with ignition on and engine not running. Startix was purchased from the Eclipse Machine Division of Bendix, and was also used on Studebaker, Pierce-Arrow, and other cars.

Auburn used the V-8's horsepower output in many of its ads for the new Cord. Oddly, they never got the figures right. The dyno chart reproduced here shows that the peak power developed by the FB engine was 117 bhp. Lycoming so informed Auburn. Whether the P.P. Willis Advertising Agency was ever told of this is unknown. But the older L-29 advertised 125 bhp, and a decrease in horsepower from a model of four years earlier was not likely to impress potential buyers. So all advertising and press releases give 125 bhp for the new Cord's horsepower rating.

Practically every car equipped with a BENDIX DRIVE is a prospect for

STARTIX

MADE BY THE MAKERS OF THE BENDIX DRIVE

Get your share of this new business!

The great majority of cars running today are equipped with the Bendix Drive. Every one in your territory represents a good profit prospect for you—SELL 'EM!

Startix is the greatest contribution to motoring safety since four wheel brakes. It completely conquers a stalled engine; keeps it running—in traffic, on hills and at crossings.

And it's a wonderful convenience—simply turn the ignition key "on" and Startix instantly, automatically starts the engine; re-starts the engine automatically if it stalls; always in control as long as the ignition is "on". No starter button necessary.

Startix is a compact, automatic electric switch that goes on the dash or engine; finished in black crystalline lacquer; easy to put on. The installation job is yours. $10 plus installation. National advertising is continually telling 26,000,000 motorists about Startix. They want it—you sell 'em!

Write us for details of our merchandising plan. Eclipse Machine Company, Department ST-42, Elmira, New York, Subsidiary of Bendix Aviation Corporation.

BENDIX STARTIX—Simplified automatic starting

BENDIX PRODUCTS

AUTOMOBILE
AVIATION-MARINE
INDUSTRIAL

```
ECLIPSE MACHINE COMPANY, Dept. ST-42, Elmira, N. Y.
  Please send me full details of your complete merchandising plan for Startix.
  Name_____
  Address_____
  City_____
```

A Bendix trade ad for Startix.

What Happened to the Showcars?

Actually, work on hand-assembled showcars continued at the same time that production tooling was being put in place. Dealers would need cars to display, if orders were to be taken. Eleven cars left for the shows at the end of October 1935. In November and December an additional 80 or more cars were assembled in Connersville.

The open show cars could not be made roadworthy. It would be months before additional bracing would permit the sale of Cord convertible coupes and phaetons that did not rattle and shake unacceptably. So all parts that could be reused were removed from the show convertibles. (One survived.)

In early 1936 a federal flood-control project was in progress in Roberts Park on the Connersville River. The bodies were hauled to the construction site in Auburn's two Federal trucks. There they were chained to posts to keep them in place, and filled with rocks to keep them down. When fill was dumped on top to complete the dam, the Cord bodies disappeared forever.

Some of the sedan bodies joined the open cars in the flood control dam. Months later, though, when the pressures had lessened, many were fitted with a new engine-transmission assembly, and with other now-available production parts. They were sold to employees at reduced prices.

A surviving price list dated December 18, 1935, gives special prices for new Cords purchased by employees of the Auburn Automobile Company, and slightly higher prices for employees of other Cord Corporation divisions. The Westchester sedan was offered to Auburn employees at $1,512.85, 25% off list. Cord Corporation workers could buy this car for $1,638.79, an 18% discount. Both prices were well below Auburn's dealer cost of $1,741.53.

Unsold show sedans stayed around the plant. They were used for parts as needed. One of them was later cannibalized for the roof of the a hard top coupe.

E. Roy Weisheit

Roy Weisheit joined Central Manufacturing Company in 1919 as a draftsman and pattern-maker. He gradually took on additional responsibilities. When E.L. Cord placed his first order for Auburn bodies in 1924, it was Weisheit who was put in charge of the Auburn account. Cord bought Central in 1928, and Roy Weisheit found himself an employee of the Auburn Automobile Company.

Over the years that Auburn and Central worked together, Weisheit and Errett Cord became close friends. According to Connersville historian Henry Blommel, Weisheit was involved in every decision that Cord made with respect to Central Manufacturing and the Auburn Automobile Company.

Weisheit was in charge of tooling for Central. He played an important part in the development of contract sales for non-automotive stampings, setting the stage for Central's continued prosperity after the automotive years ended. Central produced stampings for appliances for Crosley and General Electric. In 1934, Weisheit negotiated the sale of the tools and dies for the 1934 Auburn to Corbitt, to be used in truck manufacture. The Weisheit-negotiated contract with Montgomery Ward was critical to the production of the Cord 810.

Dies to stamp out Cord 810 body parts were ordered by Central under Weisheit's supervision. After production ended, it was Weisheit who arranged the sale of the production buildings in Auburn to Warner Gear in 1937, and the Auburn showroom to Dallas Winslow in 1938. Negotiations with Hupp for the Cord tooling in 1938 were handled by him as well.

As assistant general manager of Auburn-Central, in 1941 Weisheit obtained huge war contracts for the manufacture of B-24 bomber wings and exhaust parts. His most important deal resulted in contracts under which the former Central Manufacturing Company built nearly one-half million jeep bodies and 200,000 jeep trailers during and after W.W.II.

Roy Weisheit died in Connersville in 1979.

E. Roy Weisheit

Demonstrating the headlights. President Roy H. Faulkner and Connersville plant superintendent Curt E. Hilkey watch, Auburn vice-president Arthur Landis cranks. (Henry Blommel)

Only detail changes were made in chassis and suspension between showcars and production cars. Delco had recently licensed from Lovejoy the patent for a new shock absorber whose valving adjusted itself, albeit crudely, to the quality of the road surface. A weight inside stiffened the suspension momentarily when a sharp bump was encountered. Delco's marketing department called these Inertia-type shocks. The Model 810 was equipped with them.

Opinion on choice of universal joints for the new Cord changed almost from day to day. E-306 #2 used Bendix inner and outer joints. After the Los Angeles trip, other prototypes were equipped with Rzeppa inner and outer joints. The Rzeppa inner joint had a sliding spline to permit each axle shaft to get longer or shorter as the suspension flexed. This required that the joint be pinned to the differential. Replacing an inner joint with this design was very time-consuming, involving removal of the grille, radiator and transmission.

So production Cord 810s used the Bendix inner joint, which could accommodate axle length changes within itself. This joint could be removed without disturbing the transmission. Outer joints on the 1936 Cords were Gear Grinding's Rzeppa units.

An early cabriolet leaves Central Manufacturing. (Henry Blommel)

The hood opened from the front, a new concept it shared with the Lincoln Zephyr. The Cord's hood was much heavier, though, and its weight was not counterbalanced by springs. Lifting a Cord hood could be a chore for those of slighter stature.

Passengers' ventilation needs were more addressed in publicity than in reality. The cowl ventilator was little more than a weather-stripped lid over a screened hole in the cowl. Water poured through if opened in the rain, and the lid leaked copiously even when shut. Other contemporary cars provided a drip pan under their cowl ventilator, which admitted air but trapped water. The Cord's clean design also dispensed with windwings, so open windows were drafty. Windwings were later offered as an accessory. Rear door glass could only be opened about four inches, because of the door shape.

Supplementary ventilation was provided by windshields that opened outward individually, propelled by large knobs on the dash. The idea was to just crack them open. Fresh air would enter the car through holes in the top of the dash just below each windshield. For more sporting customers, the windshields would open out to a nearly horizontal position.

The car's shape, and careful attention to windlacing around the doors, dramatically reduced wind noise. It was this very quiet that made other noises so audible, from body squeaks to universal joint knocks.

Have You Driven a Cord, Lately?

Auburn was familiar with the Ford V-8, and did some comparisons with their new Lycoming V-8. The Cord engine displaced 288.6 cubic inches, and produced 117 bhp. (The engineers used the figure provided by Lycoming, not by their ad agency!) The Ford displaced 220.99 cubic inches and developed 90 bhp. The Cord engine weighed 575 pounds, the Ford 483. That meant that the Cord V-8 produced one horsepower for every 4.91 pounds of engine weight; the Ford weighed 5.37 pounds per horse, nearly 10% more. The Cord squeezed one horsepower out of every 1.99 cubic inches displacement. Ford's V-8 required 2.19 ci, about 10% less efficient.

Top: Twenty years later, Mrs. Fred Reesch of Auburn was still driving the show car she and her Auburn employee husband had purchased in 1935. (Randy Ema)

It was to be expected that the first car off the line would have shortcomings, and it did. Nearly all were of a minor nature. Small parts had been left off, and some components had not been properly tightened. Universal joints were leaking already, with only 385 miles on the odometer. A hopeful note was sounded: "Inner joints leak a bit . . . but that will probably disappear after some time." (It didn't.) The steering column location didn't quite match the hole in the firewall. The fuel pump was noisy, and some vibration was heard in floor panels. Still, the new Cord was a remarkable accomplishment for such a radical departure from the automotive norm, produced under such horrendous pressures of time and economics.

In February 1936, more than one hundred service managers from Auburn-Cord dealers around the country attended a two-day service course in Connersville. The men were taken through the Central plant, and saw Cord sedans coming off the line. A.H. McInnis, Auburn's supervisor of service, ran the school. Herb Snow lectured; so did Stanley Thomas, now chief engineer of Auburn. At the closing banquet, Roy

The Transmission Interlock

One of the complaints expressed by the engineers who drove E306 #2 to California was that the transmission slipped from second or third gear back to neutral. While occasional slipping out of gear was not unique to the Cord in the 1930s, it was occurring far too frequently, and was an intolerable nuisance on an expensive car.

The synchronizers on the Cord worked by moving a bronze ring attached to the synchro unit into contact with a matching steel cone on the gear. When the two parts were brought to the same speed, a sliding ring with internal splines slid smoothly over matching splines on the gear which locked gear, synchronizer and shaft into a single driving unit. All of these sliding devices had some play in them. Vibration created a tendency for the locking splines to walk out of engagement. The high helix angle of the gears created additional thrust forces. In the Cord, the too-long mainshaft bent under load and further contributed to the problem. As the shaft bent, the contact between the bronze synchro ring and steel cone was broken, reinforcing the tendency to disengage.

A brute force solution was designed to solve the problem without major redesign of the gearbox. It was developed in November 1935, and was to be installed in the first production transmissions. We're not sure exactly how this device was supposed to work, but it involved using engine vacuum to keep the synchros from moving. It apparently did not do the job and was abandoned, leaving only fragmentary design drawings behind. A mechanical linkage was devised that would keep the synchronizers in place by locking the shift forks when the clutch pedal was up. All of the factory's manpower appears to have been engaged in getting the first cars out the door, because the new interlock wasn't available to be installed until June 1936. By this time nearly 1,000 Cords had already been built and shipped to dealers; most had been delivered to customers. Cars still in stock at the factory were retrofitted with the interlock. Kits were shipped to dealers for installation in the field. But for a full five months, most new Cords slipped annoyingly out of gear. Long after this problem was solved, that glitch haunted the Cord's image.

Faulkner embroidered the facts a bit:

". . . we have spent many extra thousands of dollars and many weeks of time making this new Cord as near faultless as a mechanical unit can be built. We have tested beyond anything, to my knowledge, that has ever been done in the automotive industry. We have demanded of our workmen an exactness almost equivalent of a watch factory, and inspection of our materials is as rigid as that on a battleship."

Awards were no doubt handed out that evening to the Auburn people in attendance for their ability to maintain straight faces.

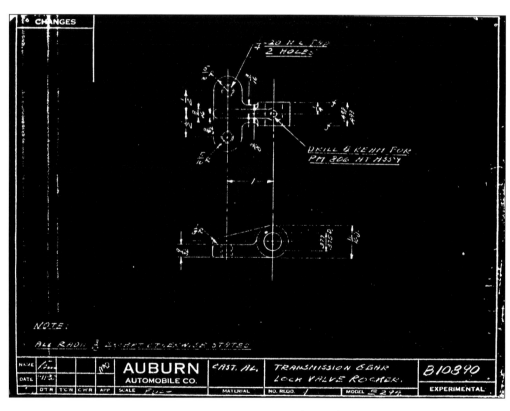

A part for the unsuccessful vacuum interlock. (Auburn-Cord-Duesenberg Museum)

The pressure was still on to produce a salable open car. The unit body of the sedans was analogous to a shoebox, and its rigidity depended on all six sides being in place. The convertibles lacked the top of the box. They rattled and shook fearfully, and doors even popped open during test runs on rough roads. A heavy cage made of welded tubing was added between the cowl and the stub frame. Although this didn't completely solve the problem, production convertible phaetons and convertible coupes finally began coming off the line in May. By that time, nearly 900 sedans had been built.

Cord sedans began arriving in February at Auburn-Cord dealerships around the country, and were soon delivered to customers. The Cord 810 was out in the world. It was now up to the dealers, their sales forces and their service people.

Model 810

June 4, 1936

Beginning with Serial No. 810-2221 the new dimensions on the bushing of the clutch and brake guide went into Production.

This dimension is a reamed diameter of the bushing of 25/22". In addition the lower 1/3" of the bushing has been removed to prevent damage to the bushing when the hose clamp used as a stop on the pedal is returned against the guide.

(This information received by telephone from McInnis June 4, 1936)

Sample page from the notes kept by the Auburn engineering departments. Original page size was 8 1/2 X 5 1/2. (Ron Irwin)

12

TO MARKET, TO MARKET

Some of E.L. Cord's major efforts, when he was rebuilding the Auburn Automobile Company, had gone into strengthening the dealer network. If a customer couldn't see and drive a car, he wasn't going to buy it. So the growth in Auburn's sales paralleled the growth in the number of dealers selling Auburns.

The dealer count peaked in 1931. It began to drop as Auburn sales did, as each fed on the other. By the time the 810 was ready for market, only 499 Auburn dealers remained in the United States, and 226 of these sold other makes as well. Chrysler Corporation had been especially active in proselytizing among disaffected Auburn dealers. Auburn had no small car, in an era when money was hard to come by. Many Auburn dealers who sold Plymouths along with Auburns and Cords soon were selling only Plymouths and Dodges, or Plymouths and DeSotos or Plymouths and Chryslers.

An 810 phaeton at a Los Angeles dealer. (Ron Irwin)

Above: A photo from The Autocar's *road test of the 810, at Brooklands Race Track.*

Right: A distributor price list, showing markups.

Auburn worked hard to help their remaining dealers sell. Sales contests were organized, coordinated by the home office in Auburn. Sales materials were relatively plentiful, for a car with such low figures. The catalog for the 810 was followed by separate brochures for the 1937 Cord, for the supercharged cars, and for the Custom Berlines. Salesmen received kits that gave them answers to every question a customer might ask, and booklets containing written testimonials from satisfied owners.

Auburn's ad agency, P.P. Willis of Chicago, had a tough job to do. Americans have always admired the new and different, but rarely rush to buy it. The new Cord was so different, from its appearance to its mechanical features, that pointing this out in advertising could have created even more sales resistance. *Advertising Age* magazine in its November 1935 issue said, "Auburn advertising will not herald the front drive from the housetops. Believing that this innovation will be regarded as too radical by some motorists, the company will endeavor, through its advertising, to sell the car as a beautiful and economical unit before going into details regarding its construction."

So Cord advertising dwelt on the fact that the Cord was different from the commonplace, without making a major issue of it. Front wheel drive was most often touted as a safety feature. Quality and

CONFIDENTIAL PRICE SHEET FOR CORD AUTOMOBILES

Effective November 2, 1935

BODY TYPE	DISTRIBUTORS' COST AT FACTORY				DEALERS' COST FROM DISTRIBUTOR'S STOCK					Equipment & Handling	List	Del. Price Retail—Freight Adver. & Tax to Add	Distrib. Profit Retail	Dealer Profit Retail
	List	Car	Equipment	Total	Car	Equipment	Handling Charge	Total	Distrib. Profit Wholesale					
Westchester Sedan 5 passenger	$1995	$1396.50	$42.00	$1438.50	$1496.25	$47.00	$15.00	$1558.25	$119.75	$75.00	$1995	$2070.00	$631.50	$511.75
Beverly Sedan with arm rests	2095	1466.50	42.00	1508.50	1571.25	47.00	15.00	1633.25	124.75	75.00	2095	2170.00	661.50	536.75
Convertible Coupe	2145	1501.50	42.00	1543.50	1608.75	47.00	15.00	1670.75	127.25	75.00	2145	2220.00	676.50	549.25
Convertible Phaeton Sedan	2195	1536.50	42.00	1578.50	1646.25	47.00	15.00	1708.25	129.75	75.00	2195	2270.00	691.50	561.75

Extra Equipment included in above Equipment Prices: Bumpers, front and rear; Spare Tire and Tube; 6 ply White Side Wall Tires all around.

	List	Distributor	Dealer
Extra for Radio	$60.00	$42.00	$45.00

ADVERTISING, LOADING, AND DELIVERY CHARGES

Distributor's Net

Advertising Charge, per car	$30.00
Loading charge, per car—truck or freight car	3.00
Half-decking—in addition to loading charge	6.00
Driveaway charge, per car	3.00
Transfer, for further shipment, of car from Connersville to Auburn or from Auburn to Connersville, per car	10.00
Customer driveaway charge to cover full retail servicing and conditioning (This is additional to regular $3.00 driveaway charge)	7.00

Extra! Extra!

In 1936, Auburn dealers were offered a new sales aid. The Reuben Donnelly Company, later famous for the telephone Yellow Pages, was making available a monthly dealer newsletter. The new publication, *Auburn-Cord News*, was directed primarily at existing customers. Donnelly made the same offer to dealers of other makes of cars. Most of the publication was generic information for motorists. A section was reserved for specific promotions for this dealer.

Auburn-Cord News was intended to provide owners and prospective customers with suggestions and tips, creating good will. Specials on seasonal service, and sales on supplies and accessories would also be announced. Copies could be delivered to subscribing dealerships each month, imprinted with the dealer's name and address. Copies could also be provided pre-addressed to the dealer's mailing list, or to a list of Auburn and Cord owners in the dealer's territory, to be supplied by Donnelly from motor vehicle registration records. Never one to miss an opportunity, Donnelly also offered to supply copies pre-addressed to local owners of 1933, 1934 and 1935 Cadillacs, Pierce-Arrows, LaSalles, Packards and Chryslers!

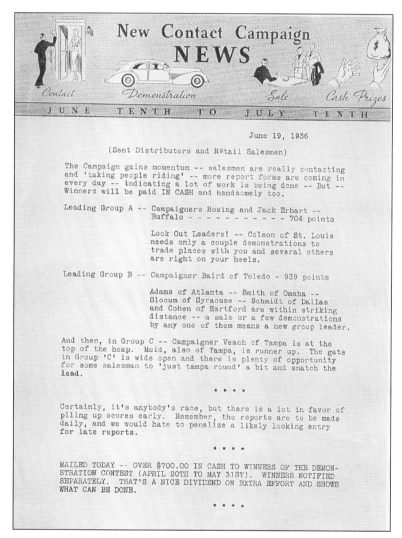

Auburn-Cord dealers ran monthly sales contests. (Ron Irwin)

safety were major themes. Willis' firm ran two distinct series of ads, each extending through both of the Cord's model years. The first simply showed a photograph of one of the Cord body styles, with the legend "Quality In Every Detail." The second series used airbrushed drawings, and considerable text describing the joys of Cord ownership. A selection of Cord advertisements appears in Appendix iii.

Willis was hampered by the fact that Auburn's advertising budget had to be shared between the Auburn car and the Cord. Between December 1935 and March 1936, for example, a total of $77,736 was spent on advertising for domestic and export sales. $35,831, or 46% of that money, was spent on advertising the Auburn car.

Still, Auburn-Cord dealers had a heck of a product to sell. A factory price sheet indicates that they could make a good profit on each car too.

The Cord's stunning appearance was enhanced by the available finishes. The company referred to the standard paint and upholstery combinations as "ensembles." They were startling for the era in their use of contrasting colors of exterior and interior. The lavish interiors of the closed cars were upholstered in English broadcloth, although leather was available as an option at extra cost in Westchester and Beverly sedans. Open cars were upholstered in leather, but the same extra charge would buy a broadcloth interior.

Leather-upholstered Beverlys used the same pattern as did the broadcloth version. Westchesters upholstered in leather were done in a unique pleated pattern. A factory memo of February 1936 indicates that Westchesters were available with this same pattern executed in broadcloth.

> **Bells and Whistles**
>
> The Cord radio dial is an integral design element of the dashboard. Even the E306s had them installed. But they were not standard equipment. Every Auburn price list to dealers spells out the radio as an optional extra-cost item. The radio could be installed at the factory or by the dealer. Sedan radios had a remote speaker with tone control mounted above the windshield. The speaker on convertible models was mounted in the radio case. Price for either radio was $60.00. If a radio was not purchased, a special round ashtray filled the holes on the right side of the dash.
> Nearly every Cord was probably delivered with a radio in place. Most dealers advertised the car's price as including the radio. Bumper guards, technically a $10.00 option, were also installed by dealers and included in the car price.

Only five color combinations were originally offered on the 810. They're described in Appendix v. In June 1936, a new two-tone scheme was offered on 810 Westchester and Beverly sedans: a Thrush Brown (Light) car with Thrush Brown (Dark) fenders.

Often repeated is the story that Roy Faulkner would do anything to sell a car, including having Cords custom painted and upholstered to customers' requests. Actually, factory memos and dealer price lists dated as early as December 1935 indicate that for an additional charge the customer could vary from the standard ensemble, and order the regular paint and upholstery colors in any combination. Further, the lists indicate that any special paint color, presumably including colors not normally available on Cords or Auburns, could be obtained as well.

Auburn dealers and salespeople were motivated and able. Working against them was the whispering campaign orchestrated by the competing dealer down the street. The Cord's well-known slippage out of gear was talked up to potential customers. So were hot weather difficulties. Those well-to-do patrons who had a sporting instinct fell in love with the sleek Cord and took the chance. The more timid bought the competition.

It didn't help that the spectacular-looking convertibles with the disappearing tops were still wretchedly flexible automobiles. Auburn's engineers continued to add reinforcement to the body and frame. In addition to the cage connecting the frame to the firewall, braces made of tubing were welded under the cowl. An array of gussets and box sections eventually stiffened the frame sufficiently to provide adequate rigidity under most driving conditions.

Top: An 810 show engine in a Los Angeles dealer's showroom. This early transmission still has the Archimedes-screw pump. (Ron Irwin)

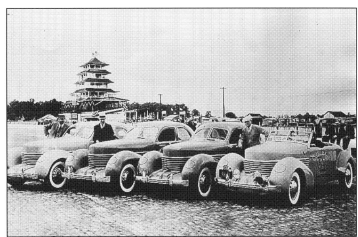

Bottom: Auburn provided Cords for the AAA observers at the 1936 Indianapolis 500 race. (Randy Ema)

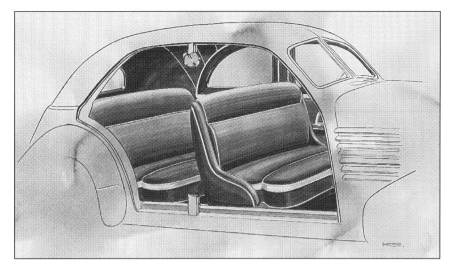

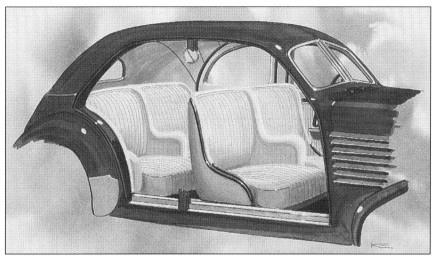

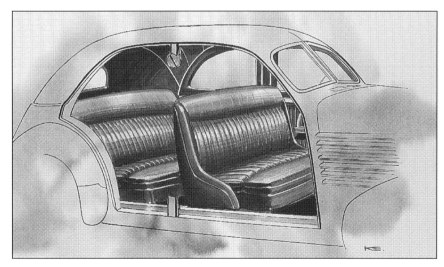

Top: The standard Westchester interior. (Ken Eberts)

Middle: The "armchair" Beverly interior for 1936. (Ken Eberts)

Bottom: The optional Westchester interior, in leather. (Ken Eberts)

Nor did it help that Auburn's sales department had suffered a loss at the top. Neil McNarby, who had been vice president of sales for ten years, died of cancer in January 1936.

Don Butler was Auburn's Car Distribution Manager, responsible for getting Cords into the hands of the regional distributors, and from them to the dealers. Auburn also had an active export department, headed by Robert Wiley. He and his employees were the equivalent of a separate car distribution department. Butler remembers that he and Wiley would meet every Monday morning to allocate the cars to be produced that week. Butler knew what was scheduled to come off the line, and would mentally set aside ten percent of the total for export. Wiley's choice of specific body styles for export didn't always agree with what was available, so they'd haggle for an hour. Butler said that he had orders for some styles, Wiley wanted other styles, so they would settle it with some give and take on both sides.

Parts lists indicate that about 10,000 parts went into a typical production Cord. To create a right hand drive model for export to countries of the British Commonwealth, fewer than 100 had to be replaced by special ones.

A substantial number of Cords did go to Commonwealth countries. Canada, England, Scotland, India, South Africa and Australia all had active Auburn-Cord dealers.

Spain and France had Auburn-Cord distributors. There were Auburn-Cord dealers in Johannesburg and in Tokyo. Auburn's European supervisor of service took a trip once to the Middle East, visiting dealerships in Cairo, Jaffa and Beirut (then spelled Beyrouth). The dealer in Austria sold Steyr cars as well. Auburn continued to seek new dealers in South America and elsewhere. Cords which are known to have been exported are so noted in the list in the appendix.

*Seeking Cord dealers in Argentina.
(Ron Irwin)*

A Star Is Reborn

Time didn't dim the Cord's glamorous image. A pair of Cords triggered special memories in *The Reincarnation of Peter Proud* (1975). Woody Allen drove a battered convertible coupe through a sidewalk cafe in *What's New, Pussycat?* (1965). In *Where The Spies Are* (1965), the film adaptation of James Leasor's novel *Passport to Oblivion,* hero Dr. Jason Love (David Niven) drives Leasor's own Cord. In the made-for-TV movie *Partners in Crime* (1978), unjustly accused Richard Jaeckel spends part of the film hiding in the back seat of an 812 phaeton driven by a judge played by Lee Grant. (Matter of fact, the car used was once Tom Mix's!) Doc Savage's car (1985) was a Cord with running boards added, to help with the action. *The Shadow* (1994) is chauffeured around in a taxi modified from an 810 Beverly. And, in the late '80s sitcom *My Two Dads,* a soft-sculpture purple 812 was a living-room centerpiece!

The glamour of the Cord was not lost on advertising agencies either, then or now. Years later Sherwin-Williams Paints and Arrow Parts used Cords in their ads. Oshkosh (b'gosh) invented a Cord Club and a meet in Montana to advertise their coveralls. TV commercials for the 1979 Dodge Magnum SE referred to its "classic Cord-type grille." In 1987 Casual Corner announced a sale with a bright yellow Cord sedan in a Florida setting. Artist Ken Eberts, whose work graces this book, won national honors for his poster for the 1987 New York Auto Show, featuring a striking view of a Cord 812. A 1989 series of ads for Chief Auto Stores used a Cord to illustrate that they carry more than just ordinary parts.

Cords stopped traffic in the 1930s. They still do today.

	List	Distributor's Net
1. Special exterior color, closed or Convertible Models	$75.00	$55.00
(Distributor to specify which standard interior to be supplied as only standard interior combination of upholstery and interior fittings can be supplied)		
2. Irregular combination of any standard exterior color and standard interior	75.00	55.00
(Cord cars are built with several standard combinations of exterior color and interior trim and any variation from these standard combinations will cost extra, as specified above)		
3. Standard Convertible Leather upholstery in Westchester or Beverly Sedans	80.00	60.00
4. Standard Sedan Broadcloth in Convertibles	80.00	60.00
(When Convertible Coupes are upholstered in Broadcloth, the rumble seat will be upholstered in Leather)		
5. Vertical Bumper Bars	10.00	7.00
6. Radio	60.00	42.00
7. Special top on Convertible models other than standard	75.00	55.00

A factory memo to dealers dated December 18, 1935. Custom paint colors and non-standard combinations of paint and upholstery were extra-cost options right from the start of production.

Many owners of the new Cord were blissfully happy with their purchase. Some cars apparently had few bugs; in other cases, owners were willing to overlook the shortcomings of their fast, gorgeous machine. Here's a selection of owner praise, culled from letters to the company, reprinted for dealer use:

"My trip home from California to Ohio was indeed a pleasant one. On the highway I got about 16 1/2 miles per gallon. I would have gotten even more, but I was going at the rate of 80 and 90 miles per hour all the way. The water temperature never exceeded 190 degrees...." "The car is a revelation. Its smooth even flow of plentiful power and speed is exhilarating; its quick acceleration is amazing. In roadability and riding comfort it is superb...." "I have driven about 650 miles in New York City and New Jersey suburbs. Every mile has been a source of pleasure. It handles the easiest of any car I have ever owned (Lincolns,Cadillacs)...." "Although it's not yet broken in— only 1,200 miles—it's become an "old friend" to the family...." "Rough roads seem transformed into boulevards as this car floats along...." "A demonstration on our ice-covered highways last February "sold" me on the new Cord...." "The Los Angeles Limited left Chicago Friday night 9:15. I left at 4:15 Saturday morning, 7 hours later. I passed that train as it was pulling into Sidney, Nebraska, 911 miles from Chicago a few minutes after 7:00 p.m. I arrived in San Bernardino California 1:20 p.m. Monday, all daylight driving. 2209 miles, 130 gallons of gasoline,

The Cord's startling modernity; the other cars on the street are only a few years old. (Henry Blommel)

changed oil every 1000 miles, less than a pint used each change. No water added, temperature never exceeded 170 degrees."

Other owners did experience problems. Auburn did its best to provide its dealer service network with factory backup, at least in this country. One of their instruments was John Hess, and he was a travelin' man. His title was Service Engineer. His job was to visit dealers, at the factory's direction, helping solve mechanical problems with the new Cord that defied the abilities of the dealer's own service department. John's territory was generally east of the Mississippi, except for the mid-Atlantic and New England states.

Randy Ema and Ron Irwin saved some of the correspondence between Hess and Auburn's service department. They provide an instructive counterpoint to the hymns sung by satisfied customers, as recounted by Auburn's publicity machine.

Repeated complaints concerned rain leaks. "The pockets on the dash simply pour water," said one woman. "Not drops, but pours on driver's feet and legs in any good rainstorm," said a man. Universal joint knocks were another common issue. Dealers were telling customers that the joints were worn, installing replacements, and having the noise recur 2,500 miles later. And, of course, there was the ever-popular slipping out of gear.

After the gear complaint, most concerns were about vapor lock in hot weather. Many of these complaints were cured by the installation of a supplementary Autopulse electric fuel pump. Dealers were provided with instructions by the factory.

A Los Angeles dealer brags about the new Cord's appeal to artists, advertising agencies and photographers. (Ron Irwin)

The Cowl Fillers

The new Cords had two lids flush with the cowl just in front of the windshields. The left one covered a fresh air vent to the interior, opened by a hand lever from inside. On the early 810s, the right one covered a pair of filler caps.

The painted caps—tan and green respectively—topped filler pipes that led to the engine. The tan one admitted water to the radiator. The green one sent oil to the crankcase. Gordon Buehrig remembered that it was Harold Ames who suggested these.

In theory, the service station attendant would not have to open the hood to check engine fluids. The dash gas gauge doubled as an oil level gauge when a button underneath it was pushed. The cowl water filler pipe was cleverly designed to be at the same height as the radiator filler neck when the car was standing level. So the level of water in the radiator was mirrored by the cowl filler.

In practice, attendants opened the Cord's hood just as they did every car. If they did use the cowl fillers they were likely as not to pour a quart of oil into the radiator, or to fill the crankcase with water!

When summer came owners realized that the exposed filler pipes passed through the inside of the car just above the front seat passenger's knees. The supplemental heating was not pleasant. Worse yet, expanding hot coolant would push past the cap in the cowl filler, and dribble out of a drain hose under the hood. The water loss promoted hot running.

By May 1936 the factory was advising dealers to remove the filler hoses, or to block them off with wood plugs and hose clamps. (A shut-off valve was considered, but all concerned agreed that it looked clumsy.) A second ventilator soon replaced the cowl fillers in production. A kit was provided to dealers to retrofit a right-side ventilator in place of the fillers.

The factory and dealers welded steel plugs into the ends of exiting cowl filler tubes where they pierced the firewall. Firewall stampings which had not yet been used had the tube section removed and a patch added. New firewalls were stamped without the tube openings.

A number of owners complained of spotty paint work. Gordon Buehrig remembered that some left-over Auburn paints had been used on the new Cords, and that the result was an uneven finish.

Hess' territory included the deep South. Even in the summer months, complaints about overheating were comparatively rare. When such complaints did occur, the factory sent Hess larger fans with instructions to follow up on whether these solved the problem. They also pointed out that water could be lost through the cowl filler, causing high temperatures. Hess was advised to plug the cowl filler tubes. The factory asked Hess to verify that a car was actually boiling, when the owner said it was overheating. If not, they suggested he move the tube in the dash temperature gauge so it would indicate ten degrees lower." "The heat indicator gauge," said Auburn Service Manager V.K. White, "is approximately ten degrees hotter than the radiator temperature." This dodge must have worked, because two months later all Cords had the gauge so adjusted on the assembly line!

Truth be told, Cords did run warmer than many contemporary cars. Part of the reason was the shape of the hood and grille, which obstructed air flow into the radiator and out of the engine compartment. (Rickenbach points out that Auburn had never shared the sheetmetal design with Lycoming, which

Added reinforcements on the stub frames of the open cars. (Ron Irwin)

The Cord and the Golden Age of Hollywood

The new Cord was above all a daringly different automobile. This uniqueness appealed especially to those who wanted to set themselves apart from the crowd. Celebrities adopted the Cord as another symbol of their special status to flaunt before an admiring public. Cords were cost-effective, too; Rolls-Royces, Duesenbergs, and some Cadillacs and Packards sold for several times the price of any Cord.

Sonja Henie, three-time Olympic ice-skating champion, drove a Cord. So did Johnnie Weissmuller, immortalized as Tarzan. Movie producer Charles Brackett, comic actor Vince Barnett and screenwriter George Bricker owned Cords too. Actor Joseph Schildkraut and boxer Max Schmeling, poles apart in every other way, were alike in their admiration of their Cords.

Prince Bernhardt of the Netherlands toured his country in his supercharged convertible coupe. (It had been given to him as an engagement gift by Queen Wilhelmina.) Amelia Earhart drove her Model 812 phaeton with pride.

Screenwriters and directors cast the glamorous Cord extensively. Placing a character in a Cord instantly identified him or her as special. Diana Gibson roared around in a convertible coupe in *Behind the Headlines* (1936). Lily Pons drove one in *Paris Honeymoon* (1939). Cords also starred in *Fifty Roads to Town* (1937), with Don Ameche and Ann Sothern; *Love Is News* (1937) starring Loretta Young and Tyrone Power; *Gangster's Boy* (1938) with Jackie Cooper; *Super Sleuth* (1937) with Jack Oakie and Ann Sothern; *Let's Get Married* (1937) with Ida Lupino and Ralph Bellamy.

Under the skin, the "Flying Wombat" featured in *Young In Heart* is really an 810 Cord. In fact, it's a unique custom creation called the Phantom Corsair, by Rust Heinz, heir to the food company fortune. Before his untimely death he planned to put it into production. (What wasn't a Cord was the car used in the "Topper" series of films; that was a Buick, with special body by Bohman and Schwartz.)

Tom Mix, the most popular cowboy star of the 1930s, had a favorite supercharged Cord 812 phaeton. He was driving at his usual high rate of speed from Tucson to Phoenix when he encountered construction barriers. The Cord swerved, skidded and overturned. Mix died of a broken neck inflicted by the steel suitcase that he carried in the back seat. A monument marks the spot today.

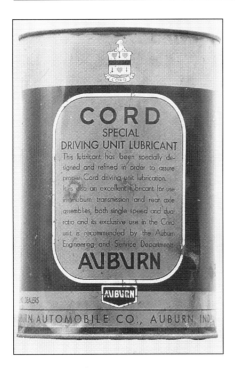

"Cord Special Driving Unit Lubricant" was sold by dealers, and came in a five-quart round can. (Ron Irwin)

did its calculations of cooling load based on the configuration of a typical Auburn.)

Auburn service departments were hampered by the fact that there had not been enough time to prepare a factory service manual for the Cord 810. Instead, "Service Bulletins" arrived every few days, to be collected in a binder. The earliest bulletins provided the service information that might have been expected in a manual. Later ones dealt with retrofits and installation of accessories.

Whenever the factory changed a specification, the information was passed on to the dealers through the service bulletins. So were corrections for bugs, and technical tips. The haste led to mistakes. A wiring diagram for the electric shift contained errors which exasperated service people. The bulletin on removing the headlight lenses gave incorrect instructions. The first service bulletin, "Installation of Cord Cylinder Head," is dated November 18, 1935, nearly three months before the cars were available. The last, "Cord Pistons," was issued on August 30, 1937, after production had ended. It was a struggle for Auburn dealers to sell and service the new Cord and the trickle of 1936 Auburns still being made, but they appear to have worked hard at it.

By October 1936, the company had changed the car's model number and year from the Model 810 of 1936 to the Model 812 of 1937. There are no records of exactly how many Model 810s were made and sold during their eight months of production. A reasonable guess is that about 1,600 810s were built, and about 1,100 sold. More than 400 unsold 810s were renumbered to be sold as 812s.

The fewer cars that were sold, the more dealers fell away, the fewer Cords were made. Within a year the number of Auburn dealers was to fall by half to 250. Only a magician could break this cycle, but Auburn was not yet entirely out of rabbits.

Top: Amelia Earhart drove a Model 812 phaeton.

Bottom: Sonja Henie, three times Olympic ice-skating champion, is believed to have owned two Cords.

Top: Love is News *(1937) starred Tyrone Power.*

Middle: The Phantom Corsair, a disguised Cord 810.

Bottom: Cowboy actor Tom Mix with his Cord.

 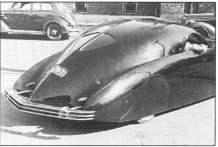

Top: Let's Get Married *(1937) with Ida Lupino and Ralph Bellamy.*

Bottom: Barton Maclane (left) wasn't happy with his Cord hung up on a rock in Angels in Exile.

Shemp Howard and Bing Crosby drove this Cord in a Paris Honeymoon.

13

THE SECOND SEASON

The United Auto Workers of America picked the inopportune month of March 1936 to organize the Auburn Automobile Company. Two months later, only four months after the Cord began to roll from the assembly line in Connersville, Auburn's management structure began to self-destruct. George Kublin had already left for a top position with GM engineering overseas. Stanley Thomas, who replaced him as Auburn's chief engineer was let go for reasons of economy. Herb Snow left to become chief engineer for Checker.

Despite all this, Roy Faulkner continued to exude optimism in his writings and talks with personnel. The layoffs and the employee grapevine said otherwise. No new Auburns had been planned for the 1937 shows. Cords were selling at what could only be called a leisurely pace. Auburn dealers were closing their doors or switching to another line of cars.

Beginning in May all engineering and administration personnel were moved to Connersville, to consolidate Auburn's management

Brochures like this one were provided by Auburn with the dealer's name imprinted on the message inside.

133

Sportsman's Convertible Coupe

Below: List of 1937 color combinations, from the salesperson's sample book. It isn't clear what the difference was between "standard" and "special" combinations. (Henry Portz)

Skiing, Hiking and Motoring

The convertible coupe was to acquire a new name, by accident. The 1937 catalog, under a photo of the two-passenger open model, referred to it as a "Sportsman's Convertible Coupe." Clearly, it was the driver who was the "Sportsman," not the car! Still, years later the term "Sportsman" began to be applied to the convertible coupe; to the dismay of purists, it seems to have stuck. For the record: there never was a Cord model called a Sportsman. In factory advertising the two-passenger open car was always called a convertible coupe. In some engineering and service materials it was sometimes called a cabriolet.

```
                    CORD
             COLOR AND UPHOLSTERY
             COMBINATIONS FOR 1937

The following selections are standard combinations on all 1937
Cord models:—

Combination No. 30......Palm Beach Tan with Maroon Broadcloth
                        (Maroon leather in Convertible models)
Combination No. 32......Black with Maroon Broadcloth
                        (Brown or Maroon leather in Convertible models)
Combination No. 41......Thrush Brown with Tan Broadcloth
                        (Brown leather in Convertible models)
Combination No. 45......Geneva Blue with Dark Blue Broadcloth
                        (Blue leather in Convertible models)
Combination No. 46......Cool Orchard Green with Green Broadcloth
                        (Green leather in Convertible models)

The following selections are special combinations, available at no
extra charge until further notice:—

Combination No. 33......Cadet Grey with Dark Blue Broadcloth
                        (Blue leather in Convertible models)
Combination No. 34......Rich Maroon with Tan Broadcloth
                        (Brown leather in Convertible models)
Combination No. 42......Clay Rust with Tan Broadcloth
                        (Brown leather in Convertible models)
Combination No. 43......Cigarette Cream with Red Leather
                        (Convertible Models only)
Combination No. 44......Ivory with Black Leather
                        (Convertible Models only)

        (Color and Upholstery specifications subject to change
                        without notice)


                   February 1, 1937
```

under one roof at Central. The move was complete by August. Gordon Buehrig and others would drive to Connersville every Monday morning, do little useful work for five days, and return to Auburn for the weekend.

It became ever clearer to those still employed, including Buehrig, that the company had no credible new projects planned. It was time for prudent folk to seek a healthier financial environment. Buehrig resigned from Auburn on August 28, 1936, without a new job offer in hand.

Alex Tremulis replaced Buehrig as head of the design department. The lower-case title is used here because the Auburn Automobile Company really didn't have a formal design department by that time in its history. The rest of the Art & Body Drafting group—Paul Reuter-Lorenzen, Dale Cosper, Vince Gardner and Dick Robinson—had left Auburn even before Buehrig did. (Cosper was in Fort Wayne as a body engineer for International Truck; Lorenzen was an illustrator for a steel company in Pittsburgh; Robinson had gone back to body drafting; Gardner worked at Budd Manufacturing in Detroit.) So, by late 1936 there were really neither department nor titles. Tremulis worked pretty much alone, perhaps with an occasional assist by Herb Newport.

The front office changed the model number of the Cord to 812 for 1937. Few mechanical changes were made for the new year. None were advertised as "new and improved," in the manner of other car manufacturers. A state-of-the-art Auto-Lite shunt-type generator, with external 3-unit regulator, was fitted. The Delco shock absorbers had dropped the inertia feature. The differential gear ratio was changed from 4.3 to 1 to 4.7 to 1. The change in ratio improved

Established 1900

AUBURN AUTOMOBILE COMPANY
Manufacturers
AUBURN, INDIANA

SERVICE DEPARTMENT

SPECIAL INSTRUCTION SHEET

NUMBER 57

AUTO-PULSE FUEL PUMP INSTALLATION.

1. Install gasoline line connector, Part No. 12S802, in place of original connector, with tapped outlet pointing toward top of car.
2. Refer to illustration showing detail location of holes to be drilled in frame for installing Auto-Pulse mounting bracket and drill two 9/32" diameter holes in the position indicated.
3. Install Auto-Pulse pump on bracket and bolt mounting bracket to bottom flanges of frame as indicated in illustration, using SF406E6 bolts, NF4E6 nuts and WA4 washers.
4. Install elbow, Part No. 12S734 in connector, Part No. 12S802, install Part No. 12S724 fitting in the inlet side of Auto-Pulse and join these fittings, using pipe assembly, Part No. 285017 and nipples, Part No. 12S704.
5. Remove carburetor elbow, Part No. 12S734 and replace with tee, Part No. 12S5780.
6. Install elbow, Part No. 12S734 in Auto-Pulse outlet.
7. Bend fuel line, Part No. 285100D6 to shape shown in illustration, fasten upper end of gas line to carburetor tee and lower end to Auto-Pulse outlet elbow.
8. Fasten gas line to dash as illustrated, using 8S64 clamps with a rubber insulator, Part No. T11033 around gas line and under clamp.
9. Run electric wire assembly, Part No. K1188 along front of dash as illustrated to rubber drain grommet where it should be carried inside body and connected to the dead side of the ignition switch.

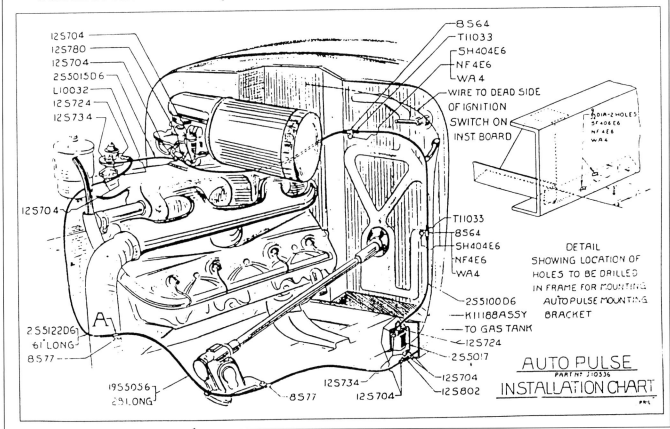

Factory instructions to dealers for installing an electric fuel pump. (Stanley Gilliland)

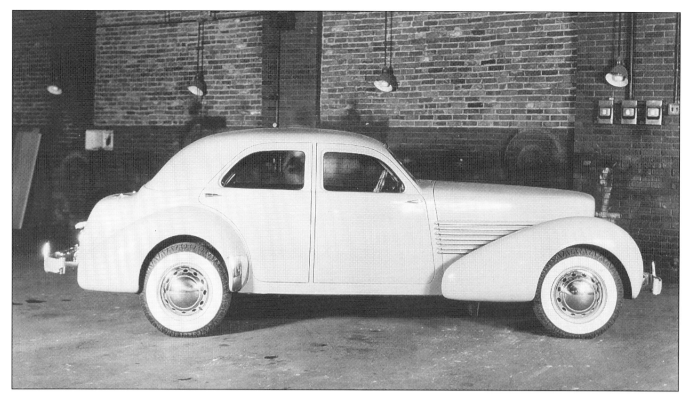

Top and bottom: An early production 812 Beverly. (Henry Blommel)

acceleration. Because the Cord's overdrive fourth gear permitted the engine to virtually loaf at cruising speeds, the lower gear ratio had relatively little effect on cruising speed or engine wear.

The Model 810 Cord, like Macbeth's nemesis Macduff, had been "ripped untimely from the womb." Early customers had the dubious privilege of serving as test drivers. Their complaints were duly forwarded to Auburn, where the shrunken engineering department struggled on bravely to perfect its product. An example was the master body fixture, which held the body panels in alignment while they were being welded together. This fixture was modified three times during the twenty months of Cord production, to improve the quality of the finished bodies.

Most changes were put in place as soon as practical, without waiting for the next model year. Springs and radiator, wheels and gauges—all were redesigned and tinkered with in a manner more appropriate to an experimental shop than to a production facility. (An order for a heavier road wheel, for example, appears to have been placed only days before the factory shut down!) In some cases the new design did not prove out, and the earlier product was reinstated. It must have driven assembly line workers crazy.

Nearly every modern front wheel drive car uses the Rzeppa patent CV joint. The Bendix-Weiss design, on the other hand, has virtually vanished. Clearly the market has found the Rzeppa design

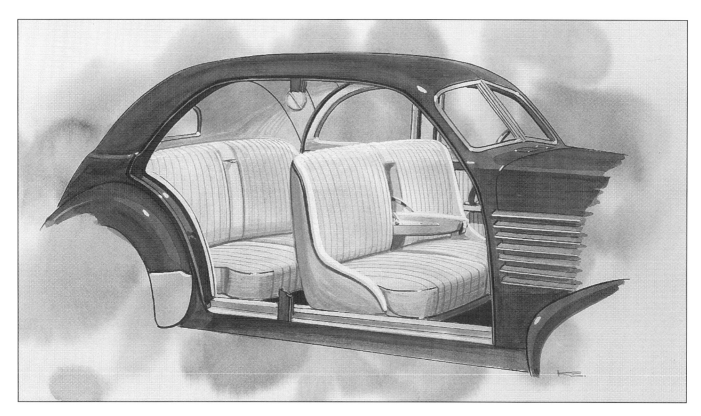

Only the Best

A catalog of "approved accessories" was issued in 1937. Some of these were manufactured by others and sold through Auburn-Cord dealers. The heater, for example, was made by EA Laboratories. Other items, like the steering wheel spinner and fender guides, were little different from what might have been purchased in an auto supply store. The factory cautioned owners, of course, not to ". . . confuse any of this merchandise with items of similar design offered at lower prices"!

superior. Still, Auburn replaced the Rzeppa outer joint with the Bendix in December of 1936. One reason often given is that the Bendix joint could be purchased more cheaply. Perhaps so. But Service Bulletin #6, issued January 15, 1937, gives instructions to dealers for ordering Bendix joints to replace Rzeppas that were less than a year old. Why?

The reason was "The Rzeppa Noise," a clicking sound that soon began to accompany Rzeppa-equipped Cords when they turned corners. The noise didn't affect the car's handling or smoothness, but it annoyed the heck out of owners.

Throughout the production period Lycoming and Auburn continued to fight the vapor lock battle that began with the FA

Top: The Beverly interior for 1937. A few Model 812s were built with this pattern in the front seat, and the 810-style armchairs in the rear. A few others were made with the armchairs in front and pull-down armrests in the rear. It's not known whether these interiors were made to customer order, or whether the factory was using up leftover seat frames. (Ken Eberts)

Bottom: The London show, 1936. (Randy Ema)

> **Paper Is Cheaper**
>
> Several possible mechanical modifications reached design stages, but were not incorporated in the production 812 models. One was a governor-controlled lockout to prevent accidental shifting into first or reverse while the car was moving. Bendix prepared a print for Auburn's approval in June 1936. Cost probably prevented adoption. It was much cheaper to hang a warning tag on the shift lever!

sample engine. For the 1937 models the fuel pump on all engines was moved from the top front of the intake manifold to the lower rear corner of the block. Wise move. The earlier location was one of the hottest spots under the hood, the later one among the coolest. With the engines being designed in Williamsport and the cars being built in Connersville, an occasional glitch could be expected. When the new engines were installed in cars, interference was found between the relocated fuel pump and the bottom edge of the firewall where it met the toeboard. The problem was resolved in typical decisive Auburn fashion—a blunt instrument was used to make just the right size dent in the sheetmetal to allow for clearance.

November came, and with it the auto show season. Once again the Cord stood out among the exhibits at the annual New York show. *TIME* magazine, in its review of that show, said the "slickest modeling in U.S. motor design can be claimed not by a new model, but by Cord introduced last year and virtually unchanged for 1937." Most other cars, truth be told, were simply face-lifts of forgettable designs.

With several major exceptions, *TIME*'s "virtually unchanged" was an accurate perception. Three of the four models which had been available in 1936—the Westchester sedan, convertible phaeton sedan, and convertible coupe—had changes in visible detail noticed even today only by the aficionado. Glove box doors had a small round keyhole, rather than a rectangular handle. The steering wheel had a thick rim with finger grips, replacing the 810's thinner smooth one. The cowl fillers for oil and water, phased out during the first months of 810 production, had been replaced by a second ventilator. That's all.

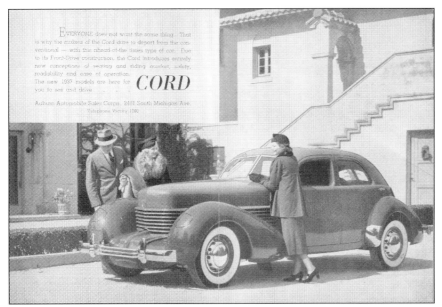

Top: The outside of another dealer brochure.

Bottom; Variations of this factory photo, taken in Los Angeles, were used on several dealer pieces. (Ron Irwin)

Opposite: Layout of the 1937 front axles, with Bendix universal joints.

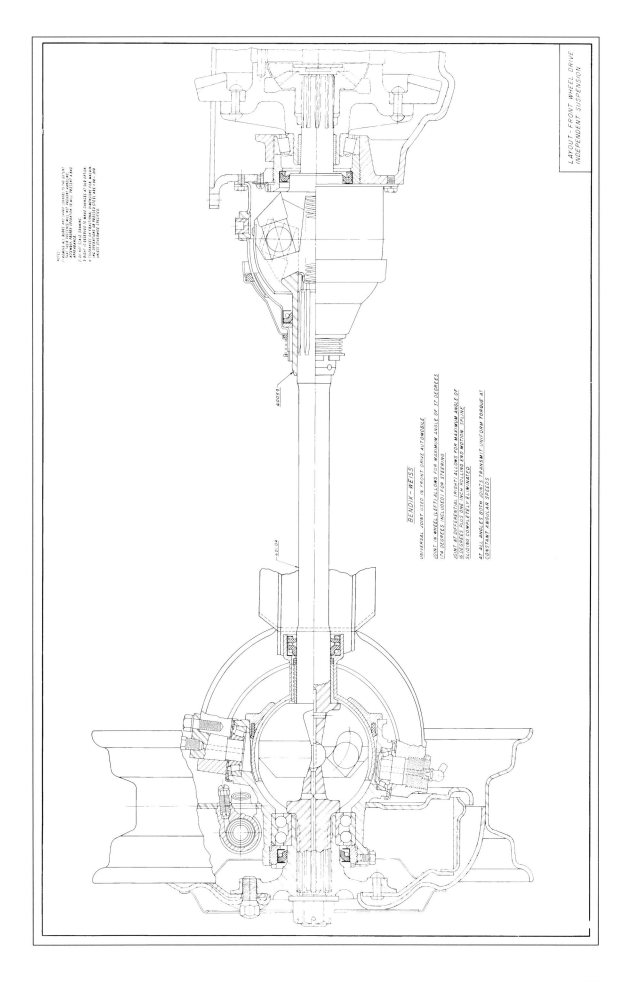

CORD FRONT DRIVE

Delivered Prices at Factory

(Advertised Prices)

Westchester Sedan	$2445.00
Beverly Sedan	2545.00
Convertible Coupe	2595.00
Convertible Sedan	2645.00
Supercharged Westchester Sedan	2860.00
Supercharged Beverly Sedan	2960.00
Supercharged Convertible Coupe	3010.00
Supercharged Convertible Sedan	3060.00
Custom Beverly	2960.00
Custom Berline	3060.00
Supercharged Custom Beverly	3375.00
Supercharged Custom Berline	3575.00

All Custom Models 132" wheelbase

Above prices include standard equipment consisting of spare tire and wheel, bumpers, bumper bars and radio.

For local delivered prices, transportation charge, state and local tax (if any) are to be added.

Auburn Automobile Company reserves the right to change prices without notice

AUBURN AUTOMOBILE CO.

Connersville, Indiana

2M-2-5-37   Printed in U. S. A.

CORD.

EFFECTIVE MARCH 1st, 1937

PRICE LIST.

39.2 h.p. 125" WHEELBASE.

WESTCHESTER SEDAN		£895
" "	Supercharged	£995
BEVERLEY SEDAN		£925
" "	Supercharged	£1025
CONVERTIBLE CABRIOLET		£935
" "	Supercharged	£1035
CONVERTIBLE PHAETON		£950
" "	Supercharged	£1050

39.2 h.p. 132" WHEELBASE.

BERLINE LIMOUSINE with Division	£1050
BEVERLEY SEDAN	£1025

RADIO is standard on all models.
All prices subject to alteration without prior notice

SOLE CONCESSIONAIRES, AUBURN, CORD, DUSENBURG.

R.S.M. [Automobiles] Ltd
26, BRUTON STREET
BERKELEY SQUARE W. I.

MAYFAIR 0283

Service: 9, Augustus Street, Regents Park, N.W.1.
Tel.: EUSTON 4007.

Left: 1937 price card distributed by dealers. A photo of the car was on the other side.

Right: Supercharged cars would not reach England until later in March 1937. This is the price list published by Auburn's London dealer.

The Beverly sedan was the only body style to undergo extensive changes for 1937. When buyers of 810 sedans complained of the scanty trunk space, the factory developed a bustle-shaped accessory trunk lid that could replace the original lid. Few were delivered. Tremulis later commented that the accessory made the beautiful sedan look like it was wearing a knapsack. He redesigned the sheet metal to integrate the new lid. (The larger trunk was welded over the opening for the original deck lid; if you peer inside, you can see the original "fastback" steel underneath.) The new shape made it necessary to move the taillamps from the trunk lid to the body. This design went into production as the Beverly sedan for 1937. Cord owners often call them "bustlebacks."

The interior of the 810 Beverly of 1936 featured luxurious armchair-type seats, and pleated upholstery. The 812 Beverly abandoned the armchairs. Herb Newport remembered that customer convenience was a factor in the decision, since the impressive arms made it impossible to slide across the seat to get out on the opposite side of the car. Also, passenger capacity was

Top: TIME *magazine took this photo at the 1937 New York Auto Show.*

Right: Cords were still pretty rare when a passerby snapped this still life on a wet day in Walcottville, Indiana.

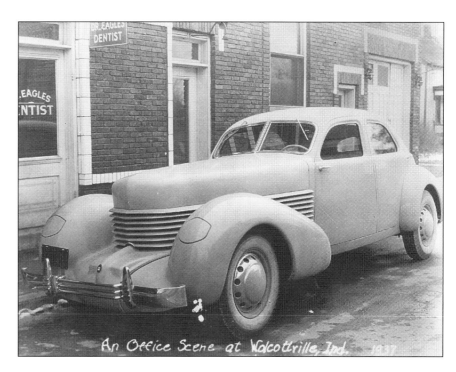

> ### The Rzeppa Universal Joints
>
> In 1935 Fred F. Miller was a draftsman working for Alfred Hans Rzeppa, whose last name is best pronounced "Sheppa." At the time of his retirement years later, Miller was a national authority on CV joints.
>
> He told me that Rzeppa joints were noisy from the start. Since pre-Cord uses were in airplane engine driveshafts and construction and industrial machinery, noise was not a major concern. It was only when the Rzeppa CV joint was installed in the quiet Cord car that noise became an issue.
>
> According to Mr. Miller, the source of problem was the ball cage. Its function was to keep the balls in their proper position. A locating pin and pilot were supposed to keep the cage positioned correctly.
>
> The locating pin did not keep the balls at exactly the correct angle in all circumstances. The "error" was not enough to affect the cars' steering or smoothness, but did cause the cage to wear rapidly. The clicking of the balls back and forth in the worn cage is the Rzeppa Noise.
>
> Miller remembers that they tried all kinds of cures. They experimented with different clearances. They tried different steels, and different heat treatments. But the clicking always started sooner or later.
>
> A previous assistant to Mr. Rzeppa, one Mr. Stuber, got the idea of generating the grooves in the outer and inner races from offset positions. The resulting cam action located the balls and cage much better than the pin had. With better ball location, cage wear was reduced.
>
> Reduced, but not eliminated. So, the Rzeppa Noise continued, although more miles could be accumulated before it started.
>
> Why are the Rzeppa-design CV joints used today so quiet? Because Mr. Rzeppa had been afraid to put too much offset in the grooves, for fear of excessive load on the outer race, and consequent breakage. Modern joints use far stronger steel, which can handle much more groove offset. The greater offset keeps the cage where it belongs. Cage wear is now negligible, and so is noise.

restricted to four and no more. For 1937, the Beverly's pleated interior used bench seats, with pull-down armrests front and rear. The redesign may have been done by Newport. Buehrig was long gone, and Newport had the needed experience with custom body interiors. Standard material remained broadcloth.

Several new standard paint and interior color combinations were added for 1937, bringing the total to ten. As before, buyers could "customize" the combinations for an extra charge.

Another change, and one that certainly would contribute to poorer sales in 1937, was an across-the-board price increase of more than twenty percent. Auburn obviously was forced to adjust prices upward to produce a profit, because they were losing money on every 810 Cord. The true costs of building each Cord may have been higher than they had estimated. Or, the sales volume, lower than expected, meant that the cost amortization for each car had to be higher. Price increases for 1937 amounted to between $450 and $500 on each model. In context, this meant adding nearly the price of a complete new Ford to that of the already high-priced Cord!

The other exceptions to *TIME's* comments were two completely new species of Cords. The Auburn car had offered an optional supercharger since 1935. It was a relatively inexpensive way to dramatically improve performance and hook customers, too. There just hadn't been time to develop the supercharger for the Cord when the 810 was being rushed to market.

Auburn had also been concerned from the start with the ability of the Cord to find a sales niche. The standard sedan was likely to be too expensive for the working man, and not expensive or large enough for the really well-to-do. The more conservative, wealthier buyer would probably want a roomier car. Again, the frantic development of the 810 hadn't permitted development of a larger Cord.

The time for both of these innovations had come.

RZEPPA
UNIVERSAL JOINT

"BUILT LIKE A BALL BEARING"

CONSTANT VELOCITY AT ALL ANGLES

AVAILABLE AS A HIGH OR LOW ANGLE JOINT IN SHAFT SIZES FROM 1" TO 2¼" FOR

FRONT DRIVES
PROPELLER SHAFTS
INDEPENDENT SPRINGING

for

Parts of Joint assembly

CARS
BUSES
TRUCKS
BOATS

The FIRST and ONLY UNIFORM VELOCITY BALL-JOINT, used by leading manufacturers for over five years.

A highly developed and proven machine element. Driving torque in either direction distributed over 6 rolling balls providing high capacity and smooth action. Ample end thrust capacity in either direction. A simple, interlocked, self-contained unit. Perfect alignment and balance. Angular capacity over 37° (74° included angle).

Working surfaces hardened and accurately ground. All parts interchangeable, designed for utmost rigidity and strength, long life and safety against break down under severe shock loads.

Send for literature and engineering details.

Special replacement assemblies for Ford trucks.

The Gear Grinding Machine Company

DETROIT   3739 Christopher Street   MICHIGAN

Gear Grinding widely advertised Rzeppa universal joints. This ad is from a June 1935 trade magazine.

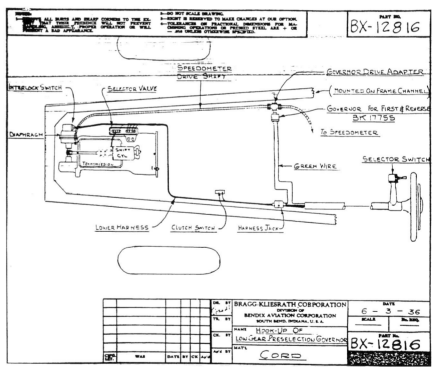

Top: This tag hung from the gear selector lever when Cords were delivered.

Bottom: Bendix's proposal for a low-reverse lockout. (Auburn-Cord-Duesenberg Museum)

I t was the *look* of the thing! That glorious, shimmering visual unity. That play of line and form so satisfying to the senses. That urge to seek new angles from which to examine and enjoy it. A viewer's tribute to the designer may be apocryphal, but is nevertheless appropriate: "It didn't look like a car. Somehow it looked like a beautiful thing that had been born and just grew up on the highway!"

For the 810 line, the convertible coupe was intended to carry four passengers; the other two would ride in the rumbleseat. The showcars were built that way. Before production of convertible coupes began in April 1936, management opted for a larger trunk instead. Gordon Buehrig remembered their thinking was that the convertible phaeton would meet the needs of those who wanted four seats.

No production 810 convertible coupes are known to have been built with rumbleseats, but there was at least one among the 812 production cars.

Convertible Coupe
810 2468F, Clay Rust
Owner: William L. Plunkett

The top on the convertible coupe and the convertible phaeton must be lowered and raised manually. The steel lid that conceals the top when it's down additionally complicates the process. Tops on Cord phaetons and convertible coupes are best dealt with by two people.

To the best recollection of those who were there, the open 1936 showcars had no tops installed. Their design had not yet been completed. The cars were exhibited with the tops down, so the steel lid hid all.

Convertible Coupe, Supercharged
812 1443F, Rich Maroon
Owner: Albert Nagele

 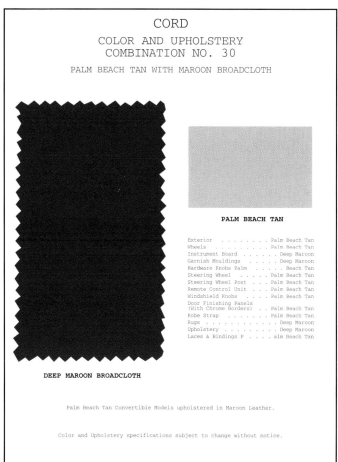

(Left) A page from the 1937 sales brochure. In the slang of the 1930s, "smart' was a synonym for "attractive."

While artist's renderings in color were used in the 1936 catalog, this is the only known use of a color photograph of a Cord in sales literature. Actually, it isn't really a color photo. It's a retouched and hand tinted version of one of the black-and-white photographs taken at Cordhaven in the fall of 1935.

Why the retoucher chose red, not a standard Cord paint color, remains a mystery.

(Right) Every dealer had a book of paint and upholstery samples from which customers could choose their color combination. Since few Auburn-Cord dealers ever had enough cars on hand to exhibit all the combinations, most choices were probably made from the books.

This sample shows one combination, as closely as ink can reproduce paint and fabric colors.

Courtesy Henry Portz

"This was a decision that management made. It probably was a result of complaints that the car was not long enough in the rear seat. . . I am sure that it was a result of criticism that the car in the price class it was in didn't have enough room in it for some people's taste." —Gordon M. Buehrig, 1966.

This is the only known example of a Custom Berline with a wheelbase of 135 inches. That's three inches longer than the stock Custom series cars. Some factory literature indicates that Berlines were built only after an order was placed. That was almost certainly the case for this one.

Custom Berline
812 10217B, Black
Owner: Kelley Auto Museum

While they don't enhance the taut lines of the Cord, these factory-installed side-mounted spare tires fit surprisingly smoothly into the pontoon fenders. The total effect is not unattractive. As bonuses, the spares provide a mounting position for dual outside rear view mirrors, a necessary accessory that the standard Cord omitted. And the missing spare does permit lots of unobstructed trunk space.

A Custom Beverly is also known to have been originally equipped with sidemounted spares. Special modifications like these are likely to have been performed elsewhere in Central's plant, after cars left the assembly line.

Convertible Phaeton Sedan
812 2156H, Palm Beach Tan
Owner: Mrs. Clarence Stanbury

(Following pages) It was easy to enter the convertible coupe through those wide front-opening doors. The spare tire stood vertically behind the seat, accessed by folding the seat backs forward. It had to be removed and replaced through the doors too. That left cavernous trunk space, by Cord standards. The upholstered sideboards concealing the spare were dropped after 810 2535F. The radio in the open cars was self-contained in the box visible under the dashboard. The speaker was mounted in the front of it. The closed cars had a separate speaker mounted in an admirable position over the center of the windshield, relatively close to the ears of the driver and front seat passenger. The ashtray under the center of the dash was an optional accessory.

Convertible Coupe
810 2468F, Clay Rust
Owner: William L. Plunkett

(Top) As with most Cord parts, there were two production versions of the hood sidescreens on the supercharged cars. The later variety is shown on this convertible coupe. It's soldered together from small brass stampings, then chromed. Auburn generally bought parts like these from outside vendors.

Convertible Coupe, Supercharged
812 31467F, Black
Owner: Ronald B. Irwin

The added length of the cowl and rear door can be seen in this example of a Custom Beverly, compared with the Westchester on the opposite page. Also visible is the bustle trunk.

Custom Beverly sedan
Geneva Blue
Photo: Auburn-Cord-Duesenberg Museum

(Top) Cord ads referred to the outside exhaust pipes as "the coat of arms of motoring royalty." A bit strong, perhaps, but they were indeed a powerful attraction for the potential buyer. From the time the first supercharged Cord rolled from the Connersville factory door to the end of production, more than half of all Cords built sported those shiny pipes. (After the factory era, it was not uncommon for outside exhausts to be added to unsupercharged cars.)

Custom Beverley Sedan, Supercharged
812 3100655, Black
Owner: Robert Gheen

The pristine lines of the original Model 810 Westchester.

Westchester Sedan
810-2087A, Cadet Gray
Owner: Josh B. Malks
Photo: Auburn Cord Duesenberg Museum

On a back road in Indiana at sunset.

While this photo was taken many years later, it is not difficult to imagine that the roads around Auburn and Connersville saw many such scenes during the heady days when Cords like this one were rolling from the assembly line.

Westchester Sedan
810 2087A, Cadet Gray
Owner: Josh B. Malks
Photo: Auburn Cord Duesenberg Museum

"To insure economical performance at higher speeds, greatest freedom of motion, to reduce wind noises, and add immeasurably to the beauty of the new Cord, we earnestly request all drivers to operate the car with headlamps retracted within the fenders during daylight driving, and bring them out only when driving conditions require lights." — Model 810 Owner's Manual

Clearly, lazy persons needed not apply to become Cord owners. The car was narrow enough so a solo driver could reach the crank for the headlight on the passenger's side of the car. The headlights went up one at a time, which could be amusing or disconcerting to approaching drivers.

A small bulb built into the lower part of the reflector provided parking lights. The effect was that of a very dim headlight.

Convertible Phaeton Sedan
810 2253H
Owner: Victor A. Kreis
Photo: Roy Kidney

T
here wasn't unanimity about the outside exhaust pipes at the 1937 auto show. Opera star Richard Bonnelli was so impressed with the pipes that he kissed Tremulis on both cheeks and called him "Maestro." Another customer was less pleased. She wanted to know "who had hung those entrails on that beautiful automobile"!

A last-minute add-on, the outside exhaust system created maintenance problems not present in the unsupercharged cars. Removing the grille for major engine service was no longer a quick job involving a half-dozen cap screws and a couple of pins; it now requiring partial dismantling of the exhaust system. Spark plugs were obscured by hot plumbing. On the open cars, with the stiffening cage structure, some underhood parts became almost inaccessible.

Actually, the pipes did have a practical benefit. Experience showed that piped cars ran marginally cooler, because the external pipes removed exhaust heat from under the hood much more effectively.

Convertible Coupe, Supercharged
812 31867F, Cadet Gray
Owner: Donald Stewart
Photo: Roy Kidney

One showcar was later purchased by Gordon Buehrig. It had a color scheme that he named Coppertone. Buehrig sent the author a colored sketch in 1959. He wrote that for the shows the body was painted sienna brown. Grille and wheels were painted a metallic copper bronze shade. (Vince Gardner believed that the original intent was to copper plate the grille and wheels.) The car shown here was later identified by Buehrig as the original Coppertone.

Components on this 810 showcar copy those used on some of the E306 prototypes, and demonstrate what they would have looked like. Grille and wheels have been copper plated to illustrate Gardner's suggestion. Inboard headlights have been fabricated, as used on the first three prototypes. The windshield has been lowered in height to resemble E306 #2. Solid hubcaps are reminiscent of that car too, as is the lack of plated rear fender gravel guards. All of the prototype cars probably sported grooved bumpers like these.

Westchester sedan
810 1021A, Special color scheme
Owner: Paul J. Bryant

A 1/32 scale model of the 810 sedan was given by Auburn distributors to most of the customers who ordered cars at the auto shows, and were disappointed by the unkept promise of delivery by Christmas 1935. There weren't enough to go around.

The model was built up of bronze investment (lost wax) castings. Windshield and window and door lines were engraved. Each bumper was a casting mounted to a strap stock bracket, and pinned to the body. Tiny door handles were separate castings. Each cast wheel had engraved crosshatched tire treads. The little car was fastened to a slab of California onyx by a machine screw passing through the base into the bottom of each tire.

Owner: Barbara Buehrig Orlando
Photos: Jordan Orlando

"We had selected about a dozen sample colors from which the five production colors were to be chosen... They did a 'market survey' by having all the stenographers observe the cars and cast their vote on their selection."
— Gordon M. Buehrig

One of the colors chosen by the stenographers was a beautiful rich maroon. Auburn's buyer got the paint manufacturer to re-pigment some leftover paint from the 1933 Auburn 12s. The result, Gordon Buehrig said later, was not as good as the color originally chosen.

Beverly sedan
810 1121S, Rich Maroon
Owner: Kathryn Buehrig
Photo: Auburn-Cord-Duesenberg Museum

14

OF LEGROOM AND SUPERCHARGERS

From the beginning, Lycoming had prepared the FB engine block to accommodate a future centrifugal supercharger, to be mounted horizontally on a new intake manifold. Bosses on which to mount the drive mechanism were incorporated into the earliest engines. So was a shoulder on the camshaft for the bevel gear that would drive the supercharger.

The Schwitzer-Cummins Corporation of Indianapolis had been supplying centrifugal superchargers to Auburn since mid-1934. The drive mechanism had been well-tested on the straight eight Auburns, and was readily adapted to the new V-8. A planetary mechanism which used rollers rather than gears stepped up the speed of the 9-inch aluminum impeller. Including the bevel gearing

Arthur Landis, Roy Faulkner, and Bert McGinnis with an early supercharged convertible. Why aren't these men smiling? (Henry Blommel)

The supercharged FC engine.

on the camshaft, the impeller whirled at 6 1/2 times crankshaft speed. The steel rollers permitted a bit of slip during sudden acceleration, easing the terrific inertial load on the bevel gears and on the timing chain. (That chain was nearly doubled in width on the supercharged engine to absorb the load of driving the blower. A new aluminum casting covered it.)

Working drawings for the supercharged V-8 were prepared soon after the production of the Model 810 started. Schwitzer-Cummins worked with Lycoming on the adaptation. While Augie Duesenberg's name has been connected with this work, according to Lycoming's Rickenbach, he was not involved. Richenbach says that Schwitzer-Cummins provided the supercharger parts; Lycoming designed the additional parts needed to fit the blower to the Cord, and had them produced by its suppliers. The aluminum impeller housing was a new casting, as was the intake manifold which replaced the manifold on the unblown, or standard V-8. These castings, like all the aluminum parts ordered by Lycoming, were delivered unfinished. Machining was done in Lycoming's own shops.

On a dynamometer chart the increased horsepower of the supercharged engine shows up mostly at the top end. That's

Indiana State Police used Cords as Safety Cars. (Randy Ema)

because a centrifugal supercharger's output is related to the square of the impeller tip speed. In theory, below about 2,000 rpm the energy required to drive the blower almost equals the increased power. In practice, in a standing-start lead-foot acceleration contest, engine revolutions rise so quickly that the supercharger's additional muscle makes itself felt every time. Rickenbach tells of the owner of a Cord 810 who visited Lycoming in 1937. He was convinced that in a short drag race his unsupercharged car could stay with, even beat, a supercharged Cord. Bill Baster was listening, and challenged the visitor to an acceleration contest against his own supercharged Cord which stood outside his office. The wager was $10. Baster collected.

Lycoming designated the supercharged engine as its series FC. The higher output of the FC engine was related to several changes beyond the addition of the blower. A Stromberg AA-25 carburetor was fitted, with a 1-1/4" throat, 1/4" larger than the EE-15. The additional height of the supercharger brought the top of the carburetor very close to the underside of the hood, so a dished, round air cleaner by Air-Maze replaced the offset design supplied by AC for the unsupercharged cars.

The FB design had an unusual firing order. All four cylinders of one bank fired, followed by the four cylinders of the other bank. No one remembers why. For the FC engine, it was felt that scavenging

Superchargers

688 superchargers were eventually sold on cars. Each was individually numbered, with a V followed by four digits. Based on the known numbers, as many as 730 superchargers may have been manufactured.

Small changes were made throughout production, as indicated by series letters cast into the aluminum housing. The first supercharger was model EA. When it was produced the fuel pump had not yet been moved to the lower rear of the block, and the intake manifold for the EA supercharger had a top-mounted fuel pump.

Top: An 812 supercharged phaeton. (Randy Ema)

Bottom: The exhaust pipe side screens are riveted together on this pre-production supercharged sedan. (Randy Ema)

Opposite: A page from *The Autocar's* road test of the supercharged 812.

would be improved by a more conventional firing order. Since the pressurized charge would create more heat, given the octane of the regular gasoline of the day, compression ratio was reduced from 6.5 to 6.32 to 1. Other under-the-hood changes, probably driven by economy, included cast iron thermostat housings and oil filler, replacing the aluminum units on the unblown engine.

The most important change, however, was the cam grind. On the first experimental FC engines, considerably more overlap was designed into the cam compared to the FB. Idle and low-speed performance were still quite satisfactory. (As we'll see, Lycoming hopped up the cam even more before production began.)

Auburn had fibbed a bit in 1936 by advertising the horsepower of the Model 810's engine at 125, eight higher than Lycoming's charts showed. The situation was strangely reversed for the supercharged engine. Advertising and press releases gave a figure of 170 bhp. In later years, slide rule engineers estimated that the 812 sedan that set AAA speed records would have had to produce 220 bhp to perform so mightily. Magazine articles and books list every figure from 130 to 225. It's even been written that changes were made to the cam grind and supercharger gearing during

March 26th, 1937.

The Autocar Road Tests

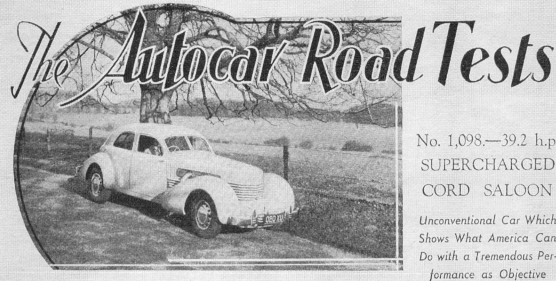

No. 1,098.—39.2 h.p. SUPERCHARGED CORD SALOON

Unconventional Car Which Shows What America Can Do with a Tremendous Performance as Objective

Normal car appearance is entirely lost in the Cord, due to the elimination of even the customary radiator grille, while the lamps are concealed in the wings and wound in and out as required. Venetian blind-type slats, which extend to the sides, admit air beneath the bonnet. Except for a small badge low down in front there is no clue to the car's identity.

IT is impossible to be indifferent to the front wheel drive Cord in its latest form. In regard, first of all, to the car's appearance, opinion is liable to be sharply divided: at first sight, people either actively like or violently dislike the unconventional shape. There is an inclination to accept the breakaway because the Cord's whole design—with front wheel drive, independent front suspension, and no normal chassis frame, for instance—is different from the ordinary, and the car does not represent just an attempt to make an otherwise orthodox vehicle look unusual. It would be less than accurate, however, to say that this car does not create a stir wherever it appears.

Æsthetic considerations apart—and they, of course, are so much a matter of personal opinion—there is no question that a terrific performance is given. This car now tested is the supercharged model, a supplementary type recently introduced. Fine and easy though the performance of the unsupercharged car proved last year to be, this present machine unquestionably excels it throughout the range of acceleration, and in top speed as well. The latter, at well above the magic 100 m.p.h. figure by stop-watch over the half-mile, is clearly altogether exceptional and remarkable.

Yet in no sense have the softness, quietness, and smoothness of an excellent type of eight-cylinder engine been impaired. The car will trickle along on its high upper ratios, and to detect any added faint sound from the supercharger when accelerating one has to listen carefully. The carburettor intake is audible to some extent in accelerating from traffic speeds, but the general effect is of an exceedingly quiet,

easy-running machine. It is superlatively good in these respects at medium and high speeds, even to as much as a genuine 80 m.p.h., wafting itself along with hardly a suggestion of mechanism working. The speeds shown by the speedometer seem quite unbelievable, yet the instrument is not seriously optimistic. It would be possible for a driver to handle the Cord and not realise that the drive is taken to the front wheels. The general quietness of running has already been stressed, and there is no trace of unusual sound resulting from the front wheel transmission, either when pulling or when on the overrun.

Really, the acceleration of this machine is tremendous, and the test figures show the point very thoroughly. Yet weather conditions were far from favourable to those tests. It is practical, usable acceleration for road work, too. By way of added interest, much can be done with the lower gears for a swift getaway, the ratios being high and the maxima they give exceptional, but more important actually is the swift pick-up on the gear corresponding to top, a ratio of 3.88 to 1. This is the gear for town work and general running around. There is a further, fourth, ratio of 2.75 to 1, a sort of super top, intended to be the fast cruising gear, though it can be kept engaged for considerable periods of time without serious loss of acceleration.

Gear changes are effected through a finger-tip lever (in effect a switch) carried by the steering column and working in a small open gate, the actual engagement being made by depressing the clutch pedal and momentarily releasing the throttle pedal, whereby a vacuum cylinder is operated. Preselection of gears is permitted. The action does not admit of a

DATA FOR THE DRIVER
39.2 H.P. SUPERCHARGED CORD SALOON.
PRICE, with four-door four-light Westchester saloon body, £995. Tax, £30.
RATING : 39.2 h.p., eight cylinders, s.v., 88.9 × 95.2 mm., 4,730 c.c.
WEIGHT, without passengers, 35 cwt. 3 qr. 12 lb.
LB. (WEIGHT) PER C.C. : 0.85.
TYRE SIZE : 6.50 × 16in. on bolt-on perforated pressed steel wheels.
LIGHTING SET : 6-volt. Automatic voltage control.
TANK CAPACITY : 17½ gallons ; approx. normal fuel consumption, 14-15 m.p.g.
TURNING CIRCLE : (L. and R.) 41ft. GROUND CLEARANCE : 9in.

ACCELERATION.

Overall gear ratios	From steady m.p.h. of		
	10 to 30	20 to 40	30 to 50
2.75 to 1	11.5 sec.	11.9 sec.	11.8 sec.
3.88 to 1	7.7 sec.	7.6 sec.	7.5 sec.
5.85 to 1	4.9 sec.	4.6 sec.	4.6 sec.
9.08 to 1	3.3 sec.	—	—

From rest to 30 m.p.h. through gears, 5.0 sec.
From rest to 50 m.p.h. through gears, 10.5 sec.
From rest to 60 m.p.h. through gears, 13.2 sec.
From rest to 70 m.p.h. through gears, 19.6 sec.
25 yards of 1 in 5 gradient from rest, 6.5 sec.

SPEED.

	m.p.h.
Mean maximum timed speed over ¼ mile	98.90
Best timed speed over ¼ mile	102.27
Speeds attainable on indirect gears—	
1st	23-34
2nd	47-60
3rd	77-88
Speed from rest up 1 in 5 Test Hill (on 1st gear)	21.64

Performance figures for acceleration and maximum speed are the means of several runs in opposite directions.

SUPER-CHARGED
CORD

production, and that the higher figures applied only to later engines.

The truth is simpler. The first tested FC engine, produced 175 bhp. Before any production engines were built, Lycoming tinkered further with the cam grind. No documentation survives of power changes, if any, resulting from the minor redesigns of the supercharged intake manifold. In sum, all production FC engines produced in the range of 186 to 195 horsepower. Curves on the original chart drawn and shaded in lead pencil show this. They, and the block-printed legend next to them, are in the hand of August Rickenbach.

There is no record of Lycoming running tests on the supercharged engine without the blower in place. Chrysler Corporation bought an FC engine in 1937, though, and ran tests on it without the blower operating. Chrysler engineers told Lycoming personnel that this engine produced 140 bhp on the dyno. (We don't know whether Chrysler used the standard or supercharged intake manifold.) The Chrysler tests help solve yet another puzzle about the supercharged Cord's output. It brings the figures more into line with the expected horsepower increase from the supercharger's six pounds of boost. An increase of 50 bhp—35%—

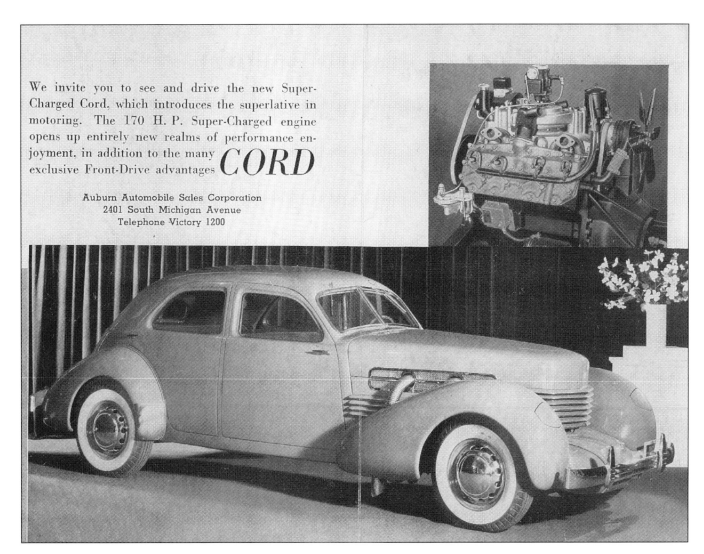

Left: Back and front of a dealer brochure.

Right: Inside copy of a dealer brochure.

over 140 unblown horses makes sense. An increase of 63% over the FB's 117 bhp doesn't. The cam made the difference.

Long-range planning was not a company strength in later days of the Auburn Automobile Company. The 1937 New York automobile show was only ten days away when Tremulis suggested to Roy Faulkner that the new supercharged cars needed exterior identification. He proposed fitting outside exhaust pipes, like those already featured on the supercharged Models 851 and 852 Auburn Speedsters and the Model SJ Duesenberg. (In theory, the outside exhausts helped remove from under the hood the additional heat created by the supercharged engine. Actually, their major function was to serve as a striking trademark for all of the Cord Corporation's supercharged cars.) Faulkner agreed, but felt, quite reasonably, that ten days was not sufficient time to prepare show models. He suggested that one be built in the coming spring for E. L.'s approval. Faulkner then left for New York and the show.

Tremulis convinced Augie Duesenberg, then heading Auburn's Experimental Garage, to try to put three prototypes together and get them to New York in time for the show. Duesenberg and his draftsmen did the design work on the exhaust system. Tremulis styled side screens for the openings where the pipes erupted from

The first production Berline. (Ron Irwin)

under the hood. The frames for screens were shaped in the Auburn shop. The screening material itself was a stock product that Tremulis ordered from Nils Melior, a New York house that supplied automotive jewelry. (They carried special radiator ornaments and similar trinkets for Duesenbergs and other cars.) At the end, Augie worked forty hours straight to complete the job. There wasn't sufficient time to ship the cars to New York, so they were driven there in fourteen hours. One was sent off low on oil, and burned a bearing en route. The two survivors were still being polished when the show opened.

There wasn't unanimity about the outside exhaust pipes at the 1937 auto show. Roy Faulkner came to the show accompanied by opera star Richard Bonnelli. The latter was so impressed with the pipes that he kissed Tremulis on both cheeks and called him "Maestro." (Faulkner apparently concealed his surprise at seeing the pipes; Tremulis says he got a raise from the boss immediately afterward.) Another customer, a gentlewoman, was less pleased. She wanted to know ". . .who had hung those entrails on that beautiful automobile"!

Still, the supercharged cars were a success with the buying public. The supercharger option was available on all models, and added $415 to the price. Cord ads for 1937 generally featured the supercharged cars, and stressed the car's high-performance nature. In advertising and factory literature, the hyphenated "Super-Charged" was used.

The Custom Cords

Gordon Buehrig later told Ron Irwin that the decision to build a longer, higher car ". . .was a result of criticism that the car in the price class it was in didn't have enough room in it for some people's taste." To reach these customers, the Cord line would have to include several larger sedans. Wheelbase would be longer for more foot room. Bodies would be higher, to provide more interior headroom, and to present a

Identifying the Custom Cord

Dan Post, in *The Classic Cord*, describes the Custom series as "improvised, rather than designed." Perhaps so, but the improvisers were a team of talented sculptors headed by Gordon Buehrig. Considering the pitfalls, the result was a brilliant success. From a distance the Custom cars are difficult to tell from their smaller fellows. Today, as then, it is often necessary even for Cord aficionados to count the grille louvers to determine whether a car is the standard or long wheelbase model. (125" cars have seven louvers; 132" wheelbase cars have eight.) To be certain that the cars were properly identified, each Custom Berline and Custom Beverly bears a small plaque near the front of the right rocker panel that clearly states "Custom Built Cord."

more impressive appearance. Interior amenities would be more luxurious, and would probably include the availability of a chauffeur-driven option.

Auburn's styling and engineering staffs were less harried than usual during the three months between the Board of Directors' decree that a brand new car was out of the question, and Faulkner's go-ahead to build prototypes. The long-wheelbase car was one of the projects on which they spent their time.

Design work on the prototype for the long-wheelbase line began in February 1935. Over the next three months the engineering department invested over 1,000 hours on this Project E296. It's listed in Auburn's work orders as "Labor on 8 cylinder 132" V type." No timesheets for Body & Art Drafting during this period are known to exist, but body styling may well have been going on simultaneously.

812-32499S, a 132" wheelbase Custom Beverly of the same height as the standard sedans. See the 7-louver hood. (Ron Irwin)

The project is specifically described as "labor," so a prototype of this car may have been built. Graphs of performance estimates indicate that the car was expected to be 400 pounds heavier than the prototype Cord sedan, and stand about 2 inches higher. These early notes suggested that the car might require 7.00-16 tires rather than the standard 6.50-16 to cope with the additional weight.

The long-wheelbase cars were later to be dubbed the Custom series. The prototypes of the series were the Berlines, with a wind-up divider window between the front and rear compartments. They were customized from standard fastback sedans in the summer of 1936.

The serial number of Berline 812 1135B indicates that the car was built as an 810, and later renumbered before sale. Its body type was C 95. (Production Berlines were body type C 103.) This car's body number, 102, shows that it was the second car of this style built. It does not have the bustle trunk of the later production Berlines. That's probably because that dubious design improvement was added by Alex Tremulis after Gordon Buehrig had left Auburn in August 1936.

The Custom series cars have a wheelbase seven inches longer than the standard 125. On production Customs, the cowl is two inches longer than on the standard cars, providing additional

A "stretched" Custom Berline, body type C 106, on an extra-long wheelbase of 135 inches. The interior view shows how the rear seat itself was moved back toward the trunk to add three more inches of footroom. (Bruce Earlin)

legroom for the driver. The balance of the additional length is in the rear seat, with five inches more legroom. On 812 1135B, the cowl is the same length as the standard cars; all seven inches of the additional length are in the rear seat. This fits well with Gordon's memory that management's original concern was for rear seat space. Rear fenders of this prototype will not interchange with the production versions.

Auburn had its hands full in 1935 getting prototypes of the 125" wheelbase sedan completed, building showcars, and trying to get production lines operating. The spring of 1936 appears to have been spent getting the open cars into production, and dealing with the bugs of the new cars. It was probably late summer of 1936 before they could turn their attention to an expanded line. Prototype Berlines were built for the November auto shows. The first production Custom Berline rolled off the line in November 1936.

Some sheet metal panels were fabricated especially for the Custom series. Others were extensively modified from parts intended for the standard wheelbase cars. The result was that few exterior body panels were interchangeable between the 125 inch wheelbase Cord sedan and the 132 inch wheelbase Custom series. Rear fenders were different. So were floorboards. The additional body height was created by higher rocker panels, creating the need for a grille with one extra louver. The taller grille permitted the use of a larger radiator. This helped provide additional cooling capacity to offset the additional weight. The higher grille also mandated special exhaust manifolds on the supercharged cars, so the pipes would exit the hood at the correct level.

More body solder, or lead, had to be used on the Custom series cars because the body had many panels with welded seams. Each seam had to be covered with lead and hand finished. Such modifications were possible only because of the low cost of skilled labor.

Berlines were upholstered in broadcloth in the rear, in a pattern different from other Cord sedans. Front seats came in a matching pattern, in either broadcloth or leather.

The Berline's rear compartment was offered in two styles. Both had wind-up glass division panels between the front and rear seats. Unsupercharged cars had a very plain rear console, with ashtray, cigar lighter and the crank for the glass divider. This car was intended to be driven by the owner or chauffeur. According to factory literature, supercharged Custom Berlines came equipped with more elaborate rear seat cabinetry, including vanity and cigarette cases, a radio speaker, and a "telephone" or intercom to the chauffeur. This deluxe divider was also available on unsupercharged cars, to special order.

The new supercharged cars, until Tremulis' last-minute intervention, were to be indistinguishable externally from the standard

Top: Auburn-Cord dealers took on franchises for other cars in 1937, to replace the defunct Auburn. This dealer adopted the Willys, seen in the background.

Bottom: A six-window Custom Beverly. The quarter window is reminiscent of the baby Duesenberg.

PICTURES TO THE EDITORS

"DON'TS"

Sirs:
I am sending you a set of pictures on "How not to get into and out of a car." The "Don'ts" this story tells are so obvious that further captions are really unnecessary.

L. E. PAULSON
Minneapolis, Minn.

THE BUSINESS OF GETTING IN (*ABOVE AND BELOW*)

THE BUSINESS OF GETTING OUT (*BELOW*)

LIFE *magazine thought that getting in and out of a Cord could be a comical experience. This piece ran in August 1937.*

The 1937 models at the London show.

models. So the Berline was probably intended to be the visual centerpiece of Auburn's display at the 1937 auto shows. As usual, it was only sixty days before the New York show when the details of the production cars were worked out.

Gordon Buehrig had left Auburn by this time. It was Alex Tremulis who styled the Berline's special interior amenities. Tremulis told of a wild 1,000-mile ride to three Midwest cities in the fall of 1936, in search of a rear seat ashtray suitable for the Berline. He was chauffeured in a Cord by Auburn test driver Ab Jenkins. The entire trip, including the shopping, took twenty hours, Jenkins driving all the way. After all that, an ashtray was chosen from the Auburn stock bins!

Tremulis also pointed out that the Berline's rear intercom and its rear radio speaker were fashioned from the bezels of the sedan's dome lights, substituting cloth for the glass. Mr. Singleton, Auburn's electrical engineer, reluctantly fashioned three-inch speakers to fit the bezels. Tremulis later chortled that this "...first breakthrough in electronic miniaturization" beat the Japanese to the punch. He recalled that the three-inch speakers actually sounded better than the standard eight-inch unit.

The other car in the Custom line was a stretched Beverly sedan with the same dimensions as the Berline. It's possible that Auburn originally intended to distinguish the sporty Beverly from the limousine-like Berline. Several examples were built with a window in the rear quarter panel, reminiscent of the baby Duesenberg. These cars carry a body type of C 101; the production Custom Beverly was C 105. High cost may have scuttled this style.

Production of the Custom Beverly began in November 1936. The interior of this car was essentially identical in design to the standard version. Upholstery material was usually broadcloth, but some leather interiors have been seen. Custom Beverlys had a cigar lighter installed in the back of the front seat, next to the ashtray. The Custom Beverly retailed for $415 more than its standard counterpart.

Some company literature indicates that the Custom series cars were built to special order. They made up about 20% of 1937 production.

The only one not satisfied with the results of the long-wheelbase project appears to have been Gordon Buehrig. He considered the Custom series to be a bastardized version of his 810 design. He told historian Ron Irwin that he was against the idea, and worked on it only because Auburn's management insisted. We don't know whether the other members of the team shared his feeling.

Buehrig's strong antipathy for the Custom series was never disguised in his later years. He was outspoken about his opinion of other abortive projects with which Auburn tried to save itself. Judge his hostility by the fact that he never voluntarily spoke of the Custom series, nor ever wrote a word about it in his books or articles.

History has dealt far more kindly with the Custom Cords. They are especially rare, and are revered as such.

15

A LITTLE BIT DIFFERENT

For a variety of purposes, some owners requested modifications on their Cords. The hardtop coupe, the bustle trunk and the LeBaron Cord resulted from some customers' special needs.

The Hardtop Coupe

One of the rarest Cord body styles was never intended for production. And it came to be, of all things, because of a man's love for his dog.

Billy Connors of Marion, Indiana, owned a string of movie theaters throughout the state. He bought a Cord convertible coupe in 1936. Connors had a Great Dane that liked to ride in the car. Much as he enjoyed the dog's company, Connors didn't want it sitting on

Left and following page, top: Views of the first Cord coupe, taken at the factory. (Henry Blommel)

Bottom: Mrs. Billy Conners and the coupe. (Jim Stover)

that nice leather upholstery. So he had a local body shop remove the convertible top and its bows, to provide space behind the seats for the dog to sit in. To maintain the Cord's appearance, the shop cut a hole in the sheet metal top lid, and lined it with foam rubber and leather. The dog sat behind the front seat, with its head sticking up through the hole.

The traveling dog was now deprived of the benefits of the windshield, and at high speeds on the highway its ears blew around. About three months later Connors went off to Auburn. He asked them to build him a two-seater closed Cord. The factory cut the top off a scrapped 1936 showcar. They welded the rear section onto a new convertible coupe just off the

In 1939 15-year-old George Peters lived in Manteca, California. Robert Stranahan Junior was dating a girl who lived down the street, and would come to pick her up in this Cord. Peters took this snapshot which happily survives as the oldest known photo of this car. (George Peters)

assembly line. The wood front bow for the convertible top remained in place to support the front of the new steel top; the wood can still be felt through the leatherette headliner. Also remaining were the pins on the windshield frame that align the top bow, and the catches that hold the front bow to the windshield. They took Connors' used convertible coupe in trade for the new hardtop coupe.

Auburn made up a new number plate for this car, to recognize the new model. The engine number and serial number were left as original, but they changed the letter suffix to "M." This was the letter suffix used for Auburn coupes, too. (Apparently the "M" stamp was missing from the set at the time the new plate was made; it's actually an upside-down "W"!) The coupe was assigned a new Central body series too: C 99. Auburn employees said that owners would sometimes come by the factory to pick up their new cars. Several saw the new coupe, and wanted one. They remembered building at least one other hardtop coupe, and possibly two.

The available evidence supports the story that this coupe and the later ones were factory modifications of convertible coupes, and not distinct body styles intended for production. On the first two coupes, for example, there is no simple way to replace a broken windshield glass.

In 1937, Auburn is believed to have created two more hardtop coupes. The first was a duplicate of Connors' 810. The second was ordered by Robert Stranahan, who was the dynamic president of Champion Spark Plug Company of Toledo, Ohio, from 1907 to 1954.

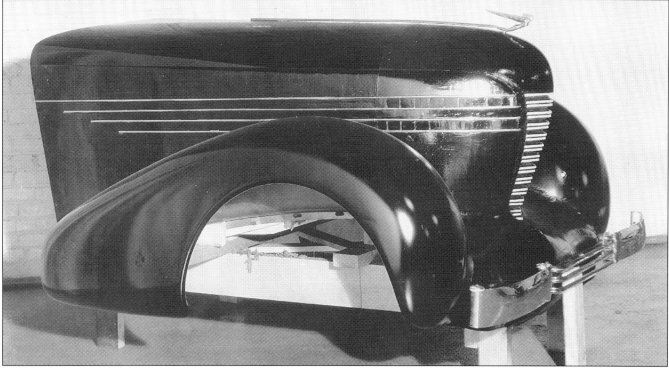

Alex Tremulis said that green clay was used to model these studies of future front end designs. (Henry Blommel)

More studies, attatched to a Cord sedan body. No one was sure what these were to be. Some nameplates say Cord, some say Auburn, some nothing at all. (Henry Blommel)

Top: A phaeton with bustle trunk. (Ron Irwin)

Bottom: This accessory trunk was offered to replace the stock 810 sedan deck lid. (Randy Ema)

Mrs. Cord's Car

The LeBaron limousine that E.L. Cord chose for his personal use was originally fitted with a Bohman and Schwartz hood and grille. Tremulis never liked it, and had one of his own design fitted before the car was delivered. This was Virginia Cord's favorite car. (She wrote "My limousine" on the back of her photo of it.) It was lost to the family for some years, and Mrs. Cord wept when she was reunited with it after its restoration by Harrah's Museum.

A master of bold marketing strategies, he pushed the gas station industry toward the concept of full service stations. He was active in civic and philanthropic endeavors, and a lover of fine automobiles.

The steel top for his car was added in the same manner as that on the earlier coupe. For this car the factory modified a stock convertible windshield to permit the glass to be replaced from the front.

This car was considerably customized, even using parts from other makes of cars. Best guess is that the work was done at the factory, but well away from the assembly line. To the new owner's specifications, chromed Auburn headlights replaced the Cord's stock disappearing units. Trim rings were used instead of the usual side screens around the external exhaust pipes. Portholes from a LaSalle were added. The steel top was covered with leather.

Opinions are mixed on whether or not this Cord is a credit to Buehrig's design. But it certainly is noticeable.

Errett and Virginia Cord's LeBaron limousine. Before delivery to them, hood and grille were changed to the style used on the LeBaron convertible sedan and convertible coupe. (National Automobile Museum)

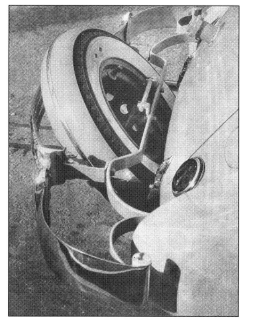

Top: A phaeton with dual sidemounted spares. (Ron Irwin)

Bottom two: The Westchester continental spare, and how it worked.

The Ten-Louver Car

One custom-bodied limousine was built, with a very high 7-passenger body. The coachbuilder is unknown; it may have been Rollston. This car used a Cord hood and grille, as modified by Alex Tremulis. To achieve the hood height needed to match the tall body, ten grille louvers were required. That meant new exhaust pipes too, and redesigned exhaust manifolds. Alex remembered "feeling like I was building a skyscraper" while stacking up the louvers.

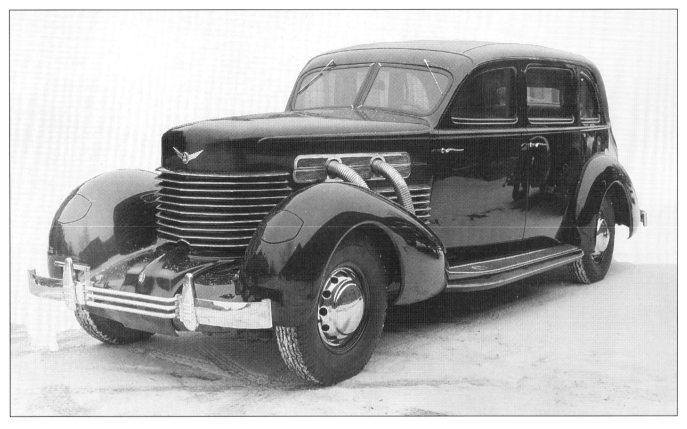

The 10-louver limousine. (Randy Ema)

The 1938 Auburns

Desperate people are inclined to take desperate action. To those who had been writing in red ink in Auburn's ledgers for years, it was obvious that the small company could not hold out much longer without some significant sales. The only car being sold by Auburn in the summer of 1937 was the Cord, and folks were not buying too many.

Why not? Was it that weird front end? Did they really associate it with a coffin? Maybe they were afraid of the front wheel drive. Or was it those rumors about transmissions and vapor locking? And maybe it was a mistake to drop the grand old name of Auburn.

Gordon Buehrig left Auburn near the end of August. The rest of the Art & Body Design people had gone before him. So Alex Tremulis was assigned the unenviable task of putting conventional front end sheet metal on the Cord 810 unit body. Perhaps that, combined with Auburn's reliable and less expensive rear drive, would permit a 1938 Auburn to be exhibited at the auto shows in November.

Henry Blommel found the photographs on pages 178 and 179, and many more, in Connersville in 1965. They are made from 8 X 10 negatives, professionally photographed for evaluation by Auburn executives. Tremulis was horrified when confronted by the sins of his youth, but manfully owned up to them. Mercifully, no one knows what became of them.

Top: Fastback sedan by LeBaron. (Henry Blommel)

Bottom: LeBaron convertible sedan. (Henry Blommel)

In Search Of More Luggage Room

Luggage space was very limited in the earliest Cord sedans and the convertible phaeton. The main response to that concern was the addition of the "bustle" trunk to the production Beverly and Custom Beverly sedans and the Custom Berline for 1937. (At least two phaetons were built with bustle trunks, too.)

To increase luggage-carrying capacity, in 1937 the factory offered a chrome luggage rack as an accessory.

The spare tire occupies much of the trunk in fastback sedans and phaetons. Removing the spare would double the usable trunk space. To this end, two 812s are known to have been built with spare tire sidemounts in the front fenders. One was a convertible phaeton, the other a Custom Beverly sedan.

Six Cords are known to have sported a factory-engineered installation of an outside spare tire on the rear. One was a Westchester sedan; five were phaetons. It is not certain whether the cars left the assembly line so equipped, or whether the outside spares were added later by a dealer. A system of brackets and pivots permitted the spare to swing down and out of the way, to allow access to the trunk. A bumper bracket of unique design was used, fabricated out of much heavier stock, to accommodate the weight of the tire. Center bumper bars were convex, and specially made for this application. The license light assembly was moved to the left rear fender to accommodate a latch arrangement at the center of the trunk lid.

LeBaron convertible coupe. No factory photos of this car exist. (Dr. John O. Baeke)

The LeBaron Customs

The idea of a smaller Duesenberg had not gone away by 1935. Stylists Alex Tremulis and Phil Wright, who both worked for Duesenberg then, were charged by Ames with the development of body styles for a new series. They were to be modern in concept but not so far-out that they would frighten away Duesenberg's traditional customers. Smaller than the Model J, and rear wheel drive, they would maintain Duesenberg's performance image by the use of a supercharged version of the Lycoming V-12 that had been used in earlier Auburns.

Tremulis and Wright developed drawings, and a contract was signed with coachbuilder LeBaron, Incorporated. Bodies were to be built of aluminum, and would not include fenders, hood, grille or bumpers. LeBaron duly delivered four body shells: a fastback sedan, a limousine with trunk, a convertible berline and a convertible coupe. All shared an identical running board design. They arrived at Central in Connersville, where they sat, keeping company with other aborted ideas. By the time the bodies arrived, neither Duesenberg nor Auburn was in a position to develop and market a new line of cars.

The Stiffer Convertible

Auburn had been struggling since the beginning with the racking of the convertible bodies. Unit bodies were a complex engineering structure. Even the Cord sedan body was a relatively primitive effort. All the doors and the trunk lid of the sedans were part of the structural box, and even these closed cars experienced some shake if everything was not properly fitted and adjusted.

Structurally, the phaetons and the convertible coupes were sedans with their tops cut off. While extra bracing under the cowl and on the stub frame helped, the problem was far from solved. It must have finally occurred to the engineers that the main frame side members were bending under the doors, since they lacked the bracing effect of the steel top.

The answer was to build up an experimental convertible with deeper, stronger frame members and rocker panels, like the Custom series was to have. The result was 812 1490F. The serial number indicates that it was modified from an 810 phaeton, probably in the fall of 1936, in the same manner that the prototype Berline was modified from an 810 Westchester. The higher rocker panels meant that an 8-louver hood was needed, again like the Custom series cars.

This must have seemed like an opportune time to experiment with styling, so Alex Tremulis remembered that he was told to "do something about that [coffin-shaped] hood." The Cord hood and grille package was a more complex design than may be casually evident; that's why it looked so good. Seen from the side, the hood and grille louvers slanted backward as they went down. Looked at from the front, the hood and louvers slanted inward as they descended. The dimensional change was not great, but it was one of the things that made the car look like it was eager to be off down the road.

Tremulis understood this, and knew what was needed to undo this look. (He admitted later that he wasn't proud of this subtle destruction of the Buehrig design.) He simply made the base of the hood-grille unit larger than the top. That made it look like a boxy lump that was about to slide down onto the transmission cover. There were echoes of ancient Renaults and Mack Bulldog trucks here, too.

This experimental car was no doubt structurally sounder than the production convertibles. Still, nobody could bring themselves to make another one.

Accessory trunk rack. (Don Howell)

Top: The experimental convertible coupe with eight louvers.

Bottom: Ab Jenkins and Augie Duesenberg built this car for Ab's personal use. (Marvin Jenkins)

187

The Eight-Stackers

Merrill M. Madsen was the Auburn-Cord dealer for Minneapolis and St. Paul. He owned a Chevrolet dealership as well. Madsen had owned and campaigned Miller racing cars in earlier years, and was a close friend of Ab Jenkins.

Madsen told the following story to A-C-D Club member Marlet "Speed" Davis in 1965: A man named White had ordered a supercharged Custom Beverly sedan at the factory, and specified some modifications because of his huge physique. Both steering column and seat positions were modified to fit his needs. White also asked for an exhaust system with four pipes on each side, instead of the standard two. The factory apparently complied.

White's car attracted a great deal of attention, and Madsen was approached by local customers who wanted to purchase their Cords with a similar exhaust arrangement. Madsen made up the complicated wooden jigs needed to hold the pipes in correct alignment while the new exhaust manifolds and headers were welded up. Special side grilles were fabricated to provide for the four outlets.

Madsen so equipped two Cords, a Beverly sedan and a convertible coupe. (He later made up jigs for a similar conversion for Cadillac V-16s.) The Cord convertible was Mrs. Madsen's personal car.

Dallas Winslow's Auburn-Cord-Duesenberg Company bought the tooling from Madsen, having received additional inquiries about similarly equipped Cords. Madsen believes that two or three more cars were built by the factory with eight outside exhaust pipes.

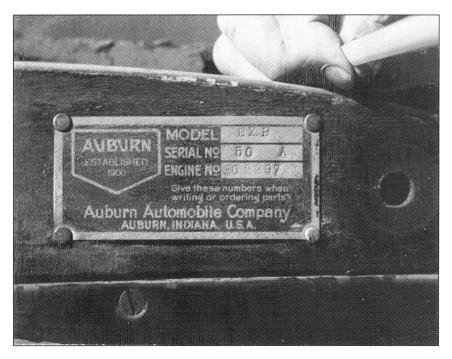

Number plate on the LeBaron limousine. This series of cars was not originally intended to be Cord-powered. (Dr. John O. Baeke)

Tremulis remembered that Errett Cord saw the bodies at Connersville during the summer of 1936, and indicated a preference for the limousine as a personal car for himself and Gee-Gee. "Put some power in that one for me," he said. The easiest way to power these cars was with the compact frame-and-drive train package of the Cord 810. And while we're doing one, thought the Auburn people, let's do them all.

On July 15, 1936, the factory Engineering Notes refer to the need for four frames for some special cars with Duesenberg bodies. After repairs were made, these frames were welded to Cord stub frames to create a wheelbase of 133 1/2 inches. On them were mounted the LeBaron bodies. They sat on top of the Auburn frame, so they were much higher than a stock Cord.

Tremulis designed several hoods and grilles for three of the cars, which were fabricated in steel, probably by Central. The fourth was designed and made by Bohman and Schwartz. Stock Cord front fenders, bumpers and hub caps were used, and Cord transmission covers were adapted. Rear fenders were stock Cord, elongated. All the cars shared a Tremulis-designed hood ornament.

The LeBaron Cords were probably completed in the fall of 1936, by which time supercharged engines were available. It's believed that all of the cars were supercharged, but no outside exhausts were used. No other cars were ever built in this line.

16
FOR THE RECORD

One of the stock car speed records held by the supercharged Cord was set about 30 days before the end of Cord production at the Connersville plant. The others were set a month after the factory shut down!

The Stevens Trophy. (All photos in this chapter except for those otherwise noted are courtesy of the Indianapolis Speedway)

These photos were probably taken the day before the run. Both supercharged Beverly sedans are free of bugs and road dirt. And, the front wheel rims don't show the inevitable stain of leaking universal joint grease symptomatic of a long drive. (Randy Ema)

Auburn badly needed a publicity coup in the spring of 1937. Auburn cars were no longer being produced, and Cords were coming off the Connersville line at a very leisurely pace. The public equated racing success with mechanical prowess and reliability, and a new record might translate into renewed customer interest at the remaining Auburn-Cord dealerships. Auburn called on its one-man "Racing Department": David Abbott "Ab" Jenkins. Ab had been employed by Auburn since 1934, and had driven nearly all of Auburn's record setting cars before and since that date.

The Stevens Trophy

The first of the two record-setting events took place in Indianapolis in June 1937. The target was "The Stevens Challenge Trophy For The Stock Closed Car 24 Hour World Record," which already had an illustrious history.

Samuel B. Stevens of Rome, New York, was the scion of a wealthy family who early entered the rich man's sport of motor racing. He raced his 60 hp Mercedes at Ormond Beach in 1904, and entered the first race for the Vanderbilt Cup later that year. He raced a more powerful Mercedes and eventually owned a string of Darracq racing cars. Stevens was one of the first members of the Society of Automotive Engineers, and an early member of the Contest Board of the American Automobile Association.

Stevens believed firmly in the benefits that racing brought to passenger car development. He also had a tongue-in-cheek attitude about the validity of advertised claims for the performance of stock cars. He wanted to offer manufacturers "an opportunity for going on record with authenticated performance." Further, he believed that the only true test of a car's merit was "to sustain top speed hour after hour." Such a run over the Indianapolis Speedway, believed Stevens, would be "even more punishing to car and driver than actual road conditions."

So he presented a trophy to the Speedway, with these provisions. Only cars manufactured in the United States were eligible (this

By the rules of the Stevens Trophy, AAA officials had to certify competing cars as factory stock. Here AAA official Tommy Thompson performs the certification procedure on Clay Rust car #2. Augie Duesenberg and Ab Jenkins wait.

despite Mr. Stevens' entire racing career having involved foreign machines). The competing vehicle must be a "fully equipped standard closed-body stock car." The holder would be the manufacturer of the car "to which is ascribed at any particular time the highest average speed for twenty-four hours over the Indianapolis Motor Speedway."

AAA would appoint the Examining Technical Committee, whose members were to be chosen exclusively from the membership of SAE. The only other requirement was that ordinary gas pump fuel be used. The winning manufacturer would hold the trophy until some other company set a new record for higher speed. A Stutz sedan was the first to try for the new record, and became the first holder of the trophy in April 1927 with an average speed of 68.44 mph. In 1931 a Marmon 16 sedan snatched it away, averaging 76.43 mph.

For Auburn's assault on the Trophy, Ab Jenkins intended to drive an Ivory-colored 1937 Cord supercharged Beverly sedan with the numeral "1" painted boldly on its sides. The Auburn team and its cars arrived during the week before the scheduled date. Pit crews set up and practiced, and AAA officials examined the Cord to certify it was indeed a stock automobile. Ab took practice runs around the Speedway oval. All was to be in readiness by Saturday night, so the

Top: At dawn on June 21, Ivory car #1 crosses the starting line to begin the 24 hour test. Ab is behind the wheel.

Bottom: Car #1 comes in for service. Augie Duesenberg is in the foreground, walking toward the right. Otto Wolfer of Firestone carries away a worn front tire.

record runs could begin early Monday morning.

By the following Tuesday afternoon, a smiling Ab Jenkins was photographed holding the Stevens Trophy as it rested on the left front fender of the record-setting Cord. Speedway and AAA officials shared the moment with him. Auburn had its new record, and the press releases would be quickly dispatched to newspapers and trade publications. Photos of the record-holding Cord appeared in contemporary publications and in enthusiast's magazines for decades to come. The car itself went on to set stock car speed records at the Bonneville Salt Flats. It was carefully preserved by a series of owners, eventually to be restored to racing trim and exhibited to universal admiration at the National Automotive Museum in Reno.

And so the matter rested for over forty years. In 1983, Auburn-Cord-Duesenberg Club member Bruce Hanson was searching for any movie film that might have survived depicting the Cord's several record runs. What he brought to light instead were 8 x 10 negatives that had rested in the Indianapolis Speedway's files since 1937, taken by photographer F.M. Kirkpatrick. They added a tantalizing mystery to the Cord/Jenkins triumph.

What the photos showed shattered the accepted notion that Jenkins' record run was simply a 24-hour tour around the Brickyard. From the photos it appeared that two supercharged Beverly sedans had been prepared for the Stevens Trophy attempt. The second car was painted a dark color and carried the number "2." The photos show it being certified as stock by AAA officials, and being fueled under the watchful eyes of Jenkins and the AAA folks.

Even more startling, the photos depict a badly damaged car #1. They also show car #2 rounding the track at speed. Was the Stevens Trophy won by car #1 or car #2? And why are there no mentions of such a dramatic mishap in the records of Auburn, or AAA, or Firestone or the Speedway?

Otto Wolfer, who was a representative of Firestone Tire and Rubber in 1937, was present at the Cord's Stevens Trophy Run. He helped me piece together what took place on that fateful weekend. Some of the photos in this chapter, never seen in print before, were made available by Marvin Jenkins and by Mr. Wolfer.

Top: Car #1 rounding a turn at speed.

Bottom: Car #2, half a lap behind, rounds the same turn. Billy Winn is at the wheel. (The photographer is panning his camera to keep the speeding car in frame. Notice how this makes the car appear to lean forward, and gives the tires an oval shape.)

The Auburn team brought two Cords to the Speedway on Thursday, June 17. Augie Duesenberg was in charge of the Auburn effort. The Duesenberg factory, only a short distance away, sent over some crew members too.

Firestone provided the tires and pit crew supervision. Firestone racing chief Waldo Stein brought Otto Wolfer and other Firestone personnel, and hired some others from the local Firestone stores. The Cords were driven at high speed in practice, to determine whether the Firestone passenger car tires would hold up on the Speedway's abrasive brick surface at these speeds.

Auburn brought two numbered cars to the Speedway for a reason. For strenuous tests like the Stevens Trophy, it is customary for two cars to make the run simultaneously. That provides a backup. If the first car should falter, the second car can continue, since it too has been circling the track since the start of the clock.

Top: Famous for presence of mind in emergencies, the skilled driver guided his shaking Cord onto the asphalt border of the infield. The car was still rolling briskly when wheel, tire and brake drum separated from the hub and spun into the front fender. The brake backing plate, front spring and front suspension ground into the pavement, and the car came to a shuddering stop.

Bottom: The Auburn pit crew, Firestone personnel and AAA officials rushed to the scene. While Augie Duesenberg examined the damage, Ab returned to the pits.

Stevens Trophy: Details Of The Run

Distance In Miles	Time	Average Speed
25.000	17.59	83.410
50.000	35.38	84.190
75.000	53.15	85.500
100.000	1:10.52	84.660
250.000	2:58.07	84.210
252.690	3:00.00	84.230
500.000	5:58.02	83.790
502.963	6:00.00	83.827
993.269	12:00.00	82.772
1000.000	12:04.41	83.170
1500.000	18:59.18	78.990
1909.851	24:00.00	79.577

Top: On his next pass, Billy Winn was flagged down and entered the pits. Car #2 took on gas; handling the hose is William Oliver, a mechanic and driver with Duesenberg, Inc. Car #2 left the pits with Ab Jenkins in the driver's seat. This procedure is not at all unusual. In Indianapolis racing as well as European Grand Prix racing, it is common practice for the lead driver to take over the next team car if his own mount breaks down. The press release issued by Auburn after the Cord won the Stevens Trophy states that "Jenkins drove his supercharged Cord for the full 24 hours, stepping out of the car for only a few seconds after he had passed the 15 hour mark." Quite so; Ab had to get out to change cars. The timing is right too. Examine the AAA Contest Board records giving the average speed at various points in the run. The average speed was never less than 82mph, except during the 12 to 19 hour segment. The change of cars dropped the average speed to 79.

Bottom: #2 passes the grandstands. The shadows indicate that it's Monday afternoon.

Top: Car #2, the Clay Rust Cord, had crossed the starting line shortly after 5:30am on Monday, June 21, 1937. By 5:30 a.m. the next day, it had been driven 1,909.851 miles, for an average speed of 79.577 mph. At no time during the 24-hour run was everyone who was involved in the effort in the pits at the same time. Firestone maintained shops at the Speedway, where many of the pit crew members had their regular jobs to work at. For this commemorative photo, never before published, all 21 members of the crew were assembled the next day. Each person in the photo was later given an 8 x 10 print. (Otto Wolfer)

Bottom: (1) Charles Merz of AAA; (2) "Snappy" Ford of AAA; (3) Tommy Thompson of AAA; (4) Sid Swaney; (5) "Pop" Myers, manager of the Indianapolis Motor Speedway; (6) Ab Jenkins, driver; (7) Billy Winn, driver; (8) Ray Gardner, mechanic; (9) Lou Webb, driver; (10) Augie Duesenberg; (11) Waldo Stein, Firestone Tires; (12) Otto Wolfer, Firestone Tires.

Race cars are driven counter-clockwise around the Indianapolis Speedway. The high speeds and banked curves put far greater pressures on the wheels on the right side of the car than would be encountered in normal driving. In a Cord, with front wheels doing both driving and steering, the stresses on the right front wheel are the greatest of all. The compact double row front wheel ball bearing that had been selected by the engineers put limits on the size and strength of the front wheel hub. At hour thirteen, about 1,100 miles into the run, the extreme pressures on this hub design triggered disaster. As Ab entered turn #1 at the southwest corner of the Speedway, the right front brake drum parted company with its hub.

The photos beginning on page 194 tell the rest of the story.

The Stock Car Records

On July 10, 1937, Ab Jenkins and Augie Duesenberg left Connersville headed for the Bonneville salt flats for an assault on the American stock car speed records. Among the crew members were Bill Oliver and Bert Updike, who would spell Jenkins in the longer runs. With them were several supercharged Beverly sedans. Marvin Jenkins remembers that a car painted Ivory and numbered "1" was among them. The crew included Waldo Stein, head of Firestone's Racing Tire Division, who brought his personal Cord.

The following day other photos were taken for publicity purposes. They may have used Car #1 because it would look better to the public. The car has been repaired, and sports a new right front fender. (The numeral "1" has been retouched too; compare the shape of the top of it with the one right after the accident.) Ab has changed into street clothes. License plate and hub caps have been re-installed, and the car is ready for its drive back to Auburn. With Ab are "Pop" Meyers of the Indianapolis Speedway, and Charlie Merz of AAA. And, of course, the Stevens Trophy. It had been won by a Cord... But not by this Cord. It would be a full sixteen years before the trophy would move on to the next winner.

How Fast Could A Cord Really Go?

Was the AAA record-holding Cord powered by a strictly stock engine? How fast could a Cord really go? Engineer Roger Huntington did some theoretical analyses in the 1950s, based on accelerometer readings and Lycoming's test charts. In his opinion it would have required at least 190 hp for the Cord to attain the speeds certified in the record runs.

We now have the charts that show that stock supercharged Cord engines produced maximum horsepower in this range. An engine carefully prepared for racing could easily produce more power without violating its strictly stock status.

The record runs described in this chapter confirm the ability of the supercharged Cord sedan to travel for substantial distances at an average top speed of 107-108 mph. Earlier, the British magazine *The Autocar* had road tested a stock supercharged Westchester sedan for their issue of March 26, 1937. They reported an average top speed of 98.9 mph, based on several runs both ways over a half-mile course. Highest speed on any one run was 102.27.

There have also been many less formal calculations of the Cord's top speed. Here are some statements by responsible people who were present when Cords traveled at higher speeds.

Lycoming's chief engineer, Bill Baster, owned a supercharged Cord sedan. He told Roger Huntington in 1950 that he had personally clocked this car in two directions on a measured mile. Top speed, he reported, was 113.

Ab Jenkins wrote in 1951 of the two Cords at Bonneville. "I drove both of those cars slightly over 120 mph—stop watch and we checked the tack [sic] against the watch so that we could check the peak, but nobody's stock tires would stay on the automobile even for ten miles at 120 mph. The treads would fly off so we didn't attempt to make a one-mile flying start record."

Horace Millhone was advertising manager for Auburn in 1937, and was present at the Bonneville record runs. He said that he took pictures of failed tire after tire, as Ab Jenkins tested different brands. Right front tires were the shortest-lived. Millhone said that the crew used to take bets on how long one would last at full throttle. Millhone confirms Jenkins' memory of top speeds during these test runs. He says that "Ab. . . hit top speeds of over 120 mph. . . on more than one occasion."

Otto Wolfer confirmed this. In tests conducted in Texas in 1937, Firestone found that few passenger tire brands were capable of sustaining speeds of one hundred miles per hour for one hundred miles. Some lasted as few as ten or twenty miles.

So it's likely that a supercharged Cord sedan could achieve top speeds greater than the records it set. How much greater, we may never know.

Bottom right: Augie and Gertrude Duesenberg drove to Utah from Indiana in the Cord 810 that the factory gave him for his personal use. (Randy Ema)

The speed runs were supervised by the American Automobile Association. Until 1979, AAA oversaw the official speed records set by American cars. The Cord competed in the "Unlimited" class.

The Cord's attempt to set new records was supposed to have taken place in late July. Storm followed storm in July and August. The drenching rains kept the salt too wet for safe driving at high speeds. Duesenberg, Updike and Oliver kept busy servicing Cords that the Salt Lake City dealer couldn't fix. One car with particularly knotty problems was sent by its owner from Idaho Falls, Idaho.

On August 13 Bert Updike was at the wheel of Waldo Stein's Cord with Stein and Augie Duesenberg aboard when it was involved in a traffic accident bad enough to wreck the car. Bert, Waldo and Augie suffered cuts and bruises. The car was sent by truck to Los Angeles for repair.

It was September before the course was dry enough for the record attempt. The whereabouts of professional photographs or movies that may have been made of this record attempt are unknown. The snapshot shown here is one of the few existing photographs. It was made during the period when the Auburn gang was waiting out the rains, and sees print here for the first time.

On September 16 and 17 the Cord set a new speed record of 107.66 miles per hour for the flying mile. The run took place on a circular course of ten-mile radius, so two way

David Abbott Jenkins

Ab Jenkins began his racing career at the age of fourteen, racing a home-built two wheel goat cart on the streets of Salt Lake City. For much of his adult life, he was a successful building contractor; racing powerful cars was then just a hobby.

By 1932, when he was 49, Ab had set records for everything from endurance runs to hill climbs to coast-to-coast driving. In that year he demonstrated the potential of the Bonneville Salt Flats by setting new stock car records with a stripped 12-cylinder Pierce-Arrow roadster. Those runs, and others the next year, showed the high speeds possible on a huge track, and brought the speed world to Bonneville. (Malcolm Campbell, George Eyston, John Cobb and Jenkins himself all set world speed records there.)

Between 1934 and 1937, the payroll records of the Auburn Automobile Company list Ab Jenkins' name in the "Racing" department. (Augie Duesenberg shows up here from time to time, also.) During these years Ab drove Auburns, Cords and farm tractors to new records at Bonneville and Indianapolis. (The Duesenberg-powered Mormon Meteor was created by Augie Duesenberg especially for Ab's attempts at world land speed records. In later incarnations, powered by a Curtis Conqueror aircraft engine, Ab drove it even faster.)

Ab's endurance became legendary. During his runs he would consume only orange juice and milk. He often drove alone for 24 hours, never leaving the driver's seat. And this, he pointed out, in cars not equipped with indoor plumbing!

In 1939 Ab was elected mayor of Salt Lake City. He made safety films shown in schools around the country, and continued to set speed records as late as 1956. A devout Mormon, he neither smoked nor drank. And in a lifetime of record-setting, he drove 1,500,000 miles without an accident.

When he died of a heart attack at 73, Ab Jenkins had set more motor racing records than any man in the world.

---

runs were not needed. A complete listing of the AAA stock car speed records set by the Cord appears in Appendix vii.

Racing puts extraordinary pressures on vehicles designed as passenger cars. The Stevens Trophy accident is an example. But let those who question the Cord's reliability note that at Bonneville Jenkins' Cord covered nearly 2,500 miles in 24 hours at an average speed of over 101 mph, including stops for gas and tires.

While the Auburn crew was waiting out the rains, the production lines in Connersville came to a halt. Someone in Auburn management must have decided that since most of the expenditures had already been made—for personnel, transportation of cars and people to Utah, food and lodging, fees for the use of the Salt Flats, fuel and oil, timers and equipment—the attempt to set new American stock car speed records would proceed. The original intent of the record runs were to provide favorable publicity and thereby boost sales. When they finally did take place the publicity might have helped dealers dispose of remaining stock.

Augie Duesenberg and the rest of the Auburn crew did not return immediately to Auburn after the new speed records were set. They remained in Salt Lake City doing further repair work on Cords at the dealerships. There was a sense of honor in those days that may seem a bit old-fashioned today. Auburn would build no more Cords. But as late as October 15 the Auburn men strove to repair the Cords they had already sold, so they would perform in a manner of which they could be proud.

Ab Jenkins and his Cord in the Salt Lake City. (Marvin Jenkins)

Car number one waits for the Salt Flats to dry. This snapshot was taken from the tower. (Marvin Jenkins)

The Final Puzzle

Ivory car #1 is shown with Ab Jenkins in photos taken after the successful American stock car speed record attempts. A car that looks like that one is shown in the snapshot from the tower at Bonneville.

AAA records indicate that the body number and engine number of the car that set the American stock car speed records were the same as those on the car that won the Stevens Trophy. We now know that the Stevens Trophy-winning car was Clay Rust in color in July, and numbered "2." Did Auburn repaint and renumber the Stevens Trophy car before it left for the salt in July? Or did Clay Rust car #2 set the new stock car speed records in its original color, with Ivory car #1 the photo stand-in again? Today the restored car is painted Ivory and emblazoned with the numeral "1." It's probably the right car. But is it the right color?

17
UNDER NEW MANAGEMENT

Many, many articles about the Cord have appeared over the years in magazines and books. Most included a mini-history of the Cord Corporation, along with comments on its demise. (Sometimes it was referred to as the "Auburn-Cord-Duesenberg Company" or the "Auburn-Cord-Duesenberg complex." Neither was strictly correct. It was Dallas Winslow who coined that name in 1938 for his parts and service business; Glenn Pray later purchased the name from Winslow.)

The closing sentences of these comments usually included words like "crumbled" or "collapsed." Some of these writings suggested that E.L. Cord personally oversaw the creation and production of the Model 810 Cord, using it as an opportunity to save his financial empire. To put this tale in perspective, we need to follow what next befell Errett Cord and his enterprises.

Cord had begun to lose interest in his automotive holdings in the early 1930s. He told Virginia during their honeymoon in 1931 that his personal role in the management of his automotive enterprises

Dallas Winslow forms the Auburn-Cord-Duesenberg Company. Winslow is third from the right. (Auburn-Cord-Duesenberg Museum)

E.L. Cord and L.B. Manning.

was over. He was not optimistic about Auburn's survival as a manufacturer of automobiles only. He turned the management of his automobile companies over to his associates, and began to pursue other interests, including aviation. (He expressed the hope to Virginia that these more profitable ventures could keep Auburn in working capital.) Cord's move to California coincided with the man's change of direction.

Errett Cord had sold his interest in the Auburn Automobile Company years before the Auburn and Cord automobiles ceased production. (Although he later bought some back, his holdings were relatively small.) Other companies controlled by the Cord Corporation, his holding company, remained in various states of financial health. Among these were American Airlines, Lycoming Manufacturing Company, New York Shipbuilding, Columbia Axle and Stinson Aircraft. The cab companies, Checker Cab and the Parmelee System, were doing particularly well. Many of the others weren't.

E.L. had been personally involved in the decisions, both corporate and mechanical, that led to the creation of the Cord L-29, and of the Duesenberg Model J. But those took place in the early years of his automotive empire-building. No records, in writing or in the memory of those who knew him, indicate that he took an active part in decisions relating to the design of the Model 810. (For most of 1934, as a matter of fact, Cord and his family lived in England, some of that time on his yacht *Virginia*.)

Cord's brief critique of E306 #2 on its visit to Los Angeles was hearkened to because the Cord Corporation controlled Auburn, but he hardly played a role in the car's development. So attributions of "desperation" or other emotions to E.L. himself with regard to decisions made at Auburn to produce the Cord 810 are, at best, conjecture. In 1935, when plans were made to produce the Cord, E.L.'s agreement was vital. And he certainly hoped that the new car could help Auburn stay solvent. Historian Randy Ema suggested, though, that his concern was mostly for the effect that this would have on the eventual sale price of his Cord Corporation stock.

The Cord Corporation in 1937 was no longer the financial giant it had been a few short years earlier. *TIME's* erstwhile "Kingdom of Cord" was teetering. The Securities and Exchange Commission was scrutinizing the corporation's stock dealings, especially those relating to Checker Cab. Cord and Morris Markin, president of Checker, signed Consent Decrees to mollify SEC. While many of the Cord Corporation's holdings were making money, Auburn and some of the other automotive enterprises were awash in a sea of red ink. It was time to get out.

Auburn-Central became American-Central in 1939. The Connersville plant carried on, with Weisheit's kitchen cabinets among other things. (Henry Blommel)

Divisions of the former Cord Corporation continued to work together. The Pak-Age-Car's rear-mounted engine was a 32-horsepower Lycoming four. Columbia made the four wheel independent suspension and transaxle. Central built the body and assembled the 15-foot long vehicle. This group of Pak-Age-Cars stands in front of the former Auburn distributor in Pittsburg, Pennsylvania in 1938. (Randy Ema)

E.L. Cord was 42 when he sold his interest in the Cord Corporation in August 1937. Most of his 30 percent of the stock went to a consortium led by financier Victor Emanuel, the rest to his old friend and associate L.B. Manning. The sale price of $2,632,000 was a tidy sum in those days. It seemed like a reasonable deal for both parties.

The new board of directors of the Cord Corporation took over on August 7, 1937. They announced that only profitable units would be continued by the new corporation. One of their first acts was to stop automobile production in Connersville. The only car being produced was the Cord 812. Don Butler, Auburn's Car Distribution Manager remembered that there were still unfinished cars on the line, and others that had been completed but not yet shipped. A skeleton force was retained in the automobile production area of Central Manufacturing to complete the cars. There were still Cords to be shipped, and still some dealers who would take them.

The Cord speed record team was in Salt Lake City when the order came to stop production. There does not appear to have been any change in their plans. As a matter of fact, in the exchange of letters and telegrams that took place afterward, there was not even a mention of the fact that Cord production lines had stopped. The men simply waited out the rains, and set new American stock car records more than a month later.

Section 77-B of the federal Bankruptcy Act protected a company while it worked out reorganization plans. The new corporate heads promptly filed for such protection for Auburn and its subsidiary, Lycoming. Supervising the case was Judge Thomas W. Slick in federal court in South Bend. (The corporation's other major properties—American Airlines, New York Shipbuilding, Vultee, Checker Cab, Columbia Axle—were apparently in no need of such protection.)

All of the Auburn Automobile Company's operations and its management personnel had been relocated to Connersville over a period of four months in 1936. The only company operation left in Auburn was the Service Parts Department. It was housed in the large building off South Wayne Street that had originally been built to assemble the 1929 Cord.

Although the Auburn Automobile Company was no longer producing cars, Lester See, Elmer Bartels and the 20 others in the Service Parts Department went right on doing their jobs. Through the fall and winter of 1937, and into 1938, they shipped out parts daily to Auburn dealers and received defective parts for exchange or credit. Their paycheck continued to come from the Auburn Automobile Company.

See remembers that executives of Warner Gear Works of Muncie, Indiana, a division of Borg-Warner Corporation, had been looking over the Auburn buildings since mid-1937. In December, Warner Gear purchased the former L-29 building for a new manufacturing plant of its own. Auburn employees quickly moved parts racks and bins into the showroom of the Auburn administration building; Warner wanted the space vacated by the day of their occupancy in March. Lester See had the forethought to apply early for a job with Warner. He left his Auburn job on March 17, 1938. He started working for Warner Gear the next day at the same kind of job in the same place. For the next 25 years Warner produced transmissions, differentials, transmission parts and axle shafts in the former Auburn factory.

Most of See's colleagues continued to work for the Auburn Automobile Company. They still shipped out parts to dealers, only now they were working out of Auburn's former showroom in the administration building.

In February 1938, the name of the former Cord Corporation was changed to Aviation and Transportation Corporation. The euphonic name was abbreviated by almost everyone as ATCO.

Dallas Winslow had been a successful automobile dealer in Flint, Michigan, in the 1920s. After selling his dealership, then dabbling in hotel ownership, Winslow began what would become a unique new career by purchasing the inventories, customer lists and good will of defunct automobile, engine and pump companies. His first purchases were the Patterson Motor Car Company of Flint, and the Flint Motor Car Company of the same city. More purchases followed. Sometimes Winslow would purchase the complete assets of these companies, including all service rights to their products. With many of these came tools and dies and blueprints. Many times he would liquidate all but the parts and service business, planning to continue such business as long as it was profitable. Winslow's General Parts Corporation in Flint continued to provide parts and service for

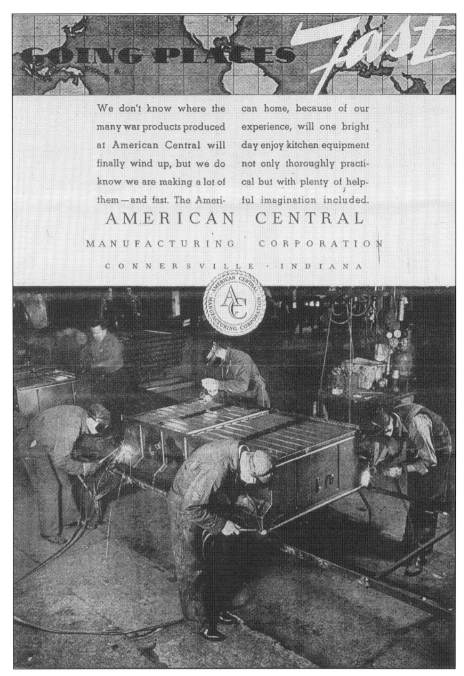

This ad appeared just before World War II began in December 1941. The production line in this photo is the same one that earlier turned out Cords. (Henry Blommel)

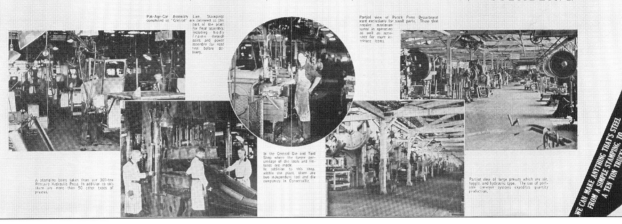

This brochure was issued in 1940. Auburn and Central were still offering their services under the same company names they had used when they built Cords together years earlier! (Ron Irwin)

these makes, even having some critical parts newly manufactured, as needed.

By the mid-1930s Winslow's headquarters were in Sidney, Ohio. He built and sold the Roto-Tiller garden tool. His Mast-Foos Manufacturing Company made replacement parts for the automobile industry. He owned Prima Manufacturing Company in Sidney, which made washing machines. His Alvork-Polk tool company in Pennsylvania produced taps and dies. McKenna Brass and Manufacturing Company made carbonated bottling equipment. And parts and service facilities for Franklin, Stearns-Knight, Locomobile, Wills St. Claire, Haynes, Briscoe, Peerless, Velie, and 20 other makes were scattered across the cities and towns of the Midwest.

In 1938, Dallas Winslow came to Auburn.

On June 15, Judge Slick in South Bend approved the sale to Winslow of the non-real-estate assets of the Auburn Automobile Company and the Auburn Automobile Sales Corporation. This purchase included the complete parts inventory for Auburn and Cord, and its service department. The price was $85,000. Also granted to Winslow was "the limited right to use the names 'Auburn' and/or 'Cord'" in business involving the sale of parts or service for the cars. The Auburn Automobile Company expressly reserved to itself the trade names "Auburn" and "Cord" with respect to the manufacture or sale in the future of automobiles by these names. Winslow also offered to purchase the remaining parts, tooling, jigs, patterns and blueprints of Duesenberg, Incorporated, from ATCO.

The Duesenberg trademark was included in the eventual sale.

At the same time Winslow bid an additional $25,000 for Auburn's art deco showroom and administration building on South Wayne Street in Auburn, and for the tools, dies and patterns belonging to the Auburn Automobile Company and located at their facility in Connersville. On July 27, Norman DeVaux of Detroit bid $45,000 for all the remaining Cord parts and assemblies at Central in Connersville, and for all the tooling, dies and patterns to make those parts. His bid also covered tooling in the hands of vendors, and blueprints and drawings for all these parts. It excluded tooling or drawings related to the mechanical parts of the car.

Fuller Motors, the Los Angeles Auburn-Cord dealer, and the radio station and transmitting tower above it. (Ron Irwin)

DeVaux got the body parts and tooling. Winslow got the administration building. Within months he had moved his inventory of parts for other cars to Auburn, and centralized his service and parts sales operations in the former Auburn administration building. His new company was called the Auburn-Cord-Duesenberg Company. Winslow's partner, H. Stanley Liddell, moved to Auburn to manage the operation.

The employees of Auburn's Service Parts Department changed jobs smoothly. One day they worked for Auburn, next day they worked for Dallas Winslow. Same workspace, new boss. Now their job included parts and service for many other makes in addition to Auburn and Cord. The most recent acquisitions would quickly become the backbone of the operation.

And what of Auburn's Central Manufacturing Company in Connersville? In addition to Auburns and Cords, Central had been producing kitchen cabinets and other non-automotive stampings since 1933. Roy Weisheit had continued to successfully seek outside contracts to use Central's stamping capacity. When the Cord line stopped, more contract work in non-automotive stampings filled the gap. He bought the patents of Russell Electric Company of Chicago, and entered the air conditioner business.

Stutz Motor Car Company of Indianapolis had staked its last hope for survival on a clever small delivery truck. The driver stood up, and the vehicle was admirably suited for stop-and-go deliveries like milk and baked goods. Stutz called it Pak-Age-Car. Stutz had purchased the rights to produce the little truck from the patent holder, Northern Motors of Chicago. Stutz was now in receivership. In September 1938, ATCO set up the Pak-Age-Car Corporation, licensed the patents from Northern and purchased the tooling from Stutz. The tooling was moved from Indianapolis to Connersville, and set up at Central. Herb Snow was then a consulting engineer for Columbia Axle, which did the final design work on the truck. Central produced the little vehicle from 1939 to 1941. They were sold through Diamond-T truck dealers.

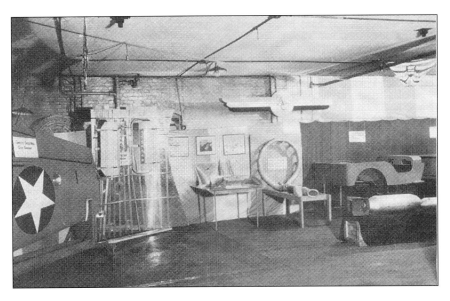
After the war was over, American-Central set up this exhibit to show off its wartime accomplishments.

Judge Slick approved Auburn's reorganization plans in 1940. The Central Manufacturing Division of the Auburn Automobile Company was separately incorporated and renamed Auburn-Central. In that year too Howard "Dutch" Darrin leased part of Central space to assemble his Packard Darrin beauties. According to Henry Blommel, they were assembled on the same line as Cords had come down until three years earlier.

In early 1941, Willys-Overland of Toledo, Ohio, won a military contract to produce a small 4 x 4 utility truck for the army. The vehicle's popular name of jeep had not yet been bestowed on it. Willys accepted bids for the construction of a lot of 1,600 jeep bodies. Auburn-Central bid and won.

By this time the former Cord Corporation had taken on the name of one of its subsidiaries, the Aviation Corporation. Avco was well positioned to be a major player in military contracts for the duration of World War II. The corporation's divisions built ships, planes and vehicles, all of which were badly needed for the war effort. Henry Blommel is convinced that E.L. Cord's influence continued to send big contracts to Connersville.

Central received a contract to build wings for the B-24 Liberator bomber. (The B-24 was a product of Consolidated-Vultee, another former holding of the Cord Corporation.) Bomber engine parts were made here too, as were fuselage sections for several different warplanes.

The biggest contract of all started with the Willys order in 1941. Over the next four years Auburn-Central would produce 445,000 jeep bodies, which would be shipped to Willys and Ford for installation on their chassis. 200,000 jeep trailer bodies would come down Auburn Central's assembly lines too, to be completed by Bantam and five other companies.

What was good for business was not good for historians. Burgeoning contract work created space shortages at Auburn-Central. In March 1941 accounting ledgers and other records of the Auburn Automobile Company that had been moved to Connersville in 1937 were carried to the town dump, there to be incinerated.

In 1942, in a burst of patriotism, Auburn-Central became American-Central. The company's war efforts were rewarded with two Army-Navy "E" pennants.

After the war, American-Central continued to supply civilian jeep bodies to Willys. Kitchen cabinets were back in production too. And "Slim" Davidson had been busy with the design of a practical consumer dishwasher. Dishwashers started down the line in 1948; American-Central underwent a name change again, this time to American Kitchens Division of Avco. Dishwashers and kitchen cabinets and appliances poured from the old Auburn Automobile Company facilities, eventually making it one of the largest

Life after the Cord Corporation

After he sold his stock in the Cord Corporation, Mr. and Mrs. Cord returned to the big house in Beverly Hills, now dubbed Cordhaven by Virginia. His business dealings continued. He took particular pride in a small kitchen appliance firm called the Chicago Electric Company. It provided uniquely designed, affordable products. Among them was an electric orange squeezer that itself became a classic design. Cord named it Sunkist.

He invested astutely in Los Angeles real estate. As a result of a defaulted loan by a large Auburn distributor Cord found himself in possession of radio station KFAC; the call letters stood for "Fuller Auburn Cord." He turned it into Los Angeles' most listened-to classical music station. He owned the Pan-Pacific Auditorium. He involved himself in gold mining, and in a fantastic scheme to use land scrip issued in the nineteenth century to control offshore oil wells. It took the federal government to stop that one.

During World War II, concerned about the vulnerability of the west coast to a Japanese invasion, E.L. built a ranch in Nevada. (Virginia told me that she didn't want to move, but did. She said that Errett didn't like people to say "no" to him. He just got sick, she said, if anyone did!)

Errett Cord was one of the United States government's "dollar-a-year" men. These were executives and industrialists who provided their expertise to the war effort at no pay. One of his closest friends was Sir William Stephenson, head of British intelligence in the Western Hemisphere during World War II. (He's best known here as "A Man Called Intrepid.") Without the knowledge of the British government then headed by appeaser Neville Chamberlain, Cord and Stephenson created an airplane factory that later poured out thousands of Spitfires when England needed them so desperately.

After the war the family moved back to the big house in Beverly Hills. The estate was sold to a developer in 1962. The night before it was razed, 500 party-goers arrived for a ball, many of them in horse-drawn carriages. Guests included luminaries of the political and entertainment worlds. Cord and his family did not attend.

In 1969 the Cords built a new permanent home in Reno, Nevada. E.L. became a Nevada state senator, and something of a political power broker.

While Errett Cord never participated in the hobby world of Cord enthusiasts, he didn't entirely snub the cars named for him. In his garage in Reno stood two Cords. The LeBaron limousine that Virginia considered her own was used for formal events. For those occasions it was chauffeur driven. Virginia told me that for personal use Errett favored new Cadillacs. Sometimes, though, when the two of them went to dinner he drove the other Cord, a supercharged Beverly. But to him they were cars, not icons. (Indeed, in later years E.L. rarely reminisced about his years in the automobile business. Randy Ema concludes that as wonderful as the cars he created are to automobile lovers, to E.L. they represented business failures. He had no patience with failure, in others or in himself.)

Errett Lobban Cord loved cars. But the manufacture of an automobile or of any other product was a business. When prospects appeared favorable, he moved to expand. When other areas appeared more lucrative or interesting, he moved in those directions. Most importantly, according to Virginia, he was always looking forward. He never looked back.

Cord was hospitalized for hemolytic anemia in 1970. He was ill with cancer when he died of a heart attack in Reno in 1974. At the time of his death, his estate was valued at over $39,000,000. His legacy to the world of automobiles is worth far, far more.

---

manufacturers of these products in the world. You'll recognize them by some of their brand names, which include White-Westinghouse and Frigidaire. In 1975, Avco sold the last of its interests in the Connersville industries.

The Auburn Automobile Company, through its divisions and successors, continued to make a mark in industry for decades after the sale of the Cord Corporation. Auburn didn't go out of business in 1937. It just stopped making Cords.

18

HUPP AND GRAHAM

Norman DeVaux was a promoter who had been in the automobile business for decades, mainly with Chevrolet and Durant. One of his companies, DeVaux-Hall Motors, had sold the short-lived DeVaux motor car in the early thirties. DeVaux contemplated the idle tooling from which the Cord body panels had been fabricated, and saw an opportunity for a rebirth of his namesake automobile.

The Victor Emmanuel group had separated the component parts of the former Cord Corporation, gathered the profitable ones that they wanted to keep and put the rest up for sale. Auburn, now in court-supervised bankruptcy, was still supplying service parts for Auburns and Cords from its Auburn location, and building kitchen cabinets at Central in Connersville.

DeVaux planned to use the Cord dies to create a low-priced car which could capitalize on the Cord's image with the public, and its still-modern appearance. To do this he needed manufacturing facilities. Joseph Graham, president of Graham-Paige, was

The proposed Hupp Junior Six. (National Automotive History Collection)

A handbuilt Hupp Skylark prototype. (Robert Fabris)

interested but Graham had no funds to spare after tooling up for its 1938 "sharknose" models.

Hupp Motor Car Company was more desperate. Its board of directors approved a plan which would make DeVaux general manager of Hupp, paying him a salary plus a royalty on each car produced. In July 1938, acting for Hupp, DeVaux made his successful bid of $45,000 for the tools, dies, jigs and fixtures used in the manufacture of the Cord. With them came some surplus body panels stamped out in 1937, but never used.

Only months later Hupp announced new lines of low priced cars for 1939, with bodies strongly resembling the late 1937 Cord. Four and six cylinder versions were planned; only the six was to reach production. The prototype of the new Junior Six used the Cord body nearly unchanged from the cowl back. From the cowl forward the sheetmetal had been shortened ten inches to accommodate the car's 115 inch wheelbase. Headlight housings were mounted inboard on the fenders, and chrome strips mimicked the Cord's louvered grille. Simpler bumpers were used, with Cord bumper guards on the front. The Cord dashboard display was reduced to a few basic instruments. Mechanically the car was pedestrian, at best. Hupp's 101 hp six-cylinder engine was used, with rear wheel drive. The archaic front suspension was by solid axle and semi-elliptic springs.

The Hupp Corsair phaeton. Only this prototype was built. (Robert Fabris)

Dealers and the public probably had mixed reactions. It was one thing to capitalize on the Cord's good looks, quite another to offer what was essentially a parody. (It's possible that Hupp never actually intended to produce the imitation Cord Junior Six, but had to throw something together quickly to maintain dealer interest.) Briggs stylist John Tjaarda, of Lincoln-Zephyr fame, was engaged to quickly create a new front end appearance. His redesign of the hood and grille was masterful. Hupp also added windwings to the front and rear door glass, to improve ventilation. Waldo Gernandt re-engineered the Cord's unit body to accept the rear drive mechanical parts. The new car was shown at the New York Auto show the following month, with a new front end and a new name. It was now the Hupp Skylark.

Hupp announced three body styles—Skylark sedan, Sportster cabriolet and Corsair phaeton—with the sedan available in several trim levels. Again the public and the press responded with enthusiasm. The prestigious American Federation of Art gave special attention to the Skylark in its annual review of automobile design. Noting that the Skylark was adapted from the Cord, itself a handsome automobile, the reviewer babbled ". . . anyone looking for first-class design will recognize improvements in this car over its more costly forerunner."

Gordon Buehrig was less impressed. He remembered being unhappy at what had been done to his Cord design, but allowed that Tjaarda had done a creditable job, given what he had to work with. (Time has come down on Gordon's side; the Skylark and the Graham Hollywood are today considered little more than grace notes to the timeless Cord composition.)

A familiar story now repeated itself. Auburn's presses, for which the Cord dies were designed, were relatively small. So the Cord was built of many pieces welded together with hand-applied lead covering the seams. This was problem enough in a relatively high priced car; it was a disaster when applied to a low priced vehicle. Problems with the multi-part roof and fender dies held up production of the first Skylarks until March 1939. Only 31 sedan prototypes had been built when Hupp's financial situation forced the cessation of production in late May. Only one Corsair convertible was made.

Re-enter Graham-Paige, which was having its own problems. Production of the shark-nose Graham had fallen to 6,000 units in 1939. Graham too now saw salvation in the Cord dies. They proposed that assembly of the Hupp Skylark be transferred to Graham's Dearborn plant. A deal was struck between DeVaux, who still owned the tooling, and between Graham and Hupp. The dies were moved to Graham's plant. Graham agreed to make improvements that would render the

> **Old Tooling Never Dies**
>
> During the 1950s a story surfaced regarding Japanese ownership of the Cord body dies. Motoring magazines repeated variations of the story. Supposedly an anonymous Cord fan had glimpsed the Cord body dies in packing crates at a Japanese port. Story endings varied. The dies were now being used to make Toyotas. They were enshrined in Japan by a wealthy Japanese art lover, who could not bring himself to destroy objects that had created such beauty. They were being stored in a warehouse in Japan, lost because they are mislabeled.
>
> The story is probably much simpler. Since Auburn used small dies to piece together large parts, its unlikely that the Cord dies would have been recognizable as such if they were seen in a crate. The dies probably became World War II scrap metal in the United States. If they were indeed sold to Japan, it would also have been as scrap. In that case, we probably received them back in 1942 in a different form.

dies more suitable for mass production. In particular, Graham would create a single-piece roof die to replace the Cord's labor-greedy multi-piece roof.

Graham and Hupp cars based on these dies would be produced on the same assembly line. Graham would gain a new low-priced car for nearly no investment. Hupp would pay a modest fee for each body built by Graham. Hupp engines, running gear and trim would be installed on those cars to be sold as Skylarks. Graham's supercharged 120 hp six-cylinder Continental-built engine would make its version of the car a standout performer. It would be called the Graham Hollywood.

A Graham Hollywood. (National Automotive History Collection)

Transferring Hupp's equipment and the Cord dies to Graham's plant took longer than expected. So did the modifications to the Graham assembly line to accommodate the new car's unit body construction. And Graham simply did not have the money for the new roof die. So it was May 1940 before Skylarks and Hollywoods were available for sale. In the interim, photos of the new cars were on display at dealerships, with badges and trim modified to suit the make of car that the dealer sold.

1940 production ended the first week in July. 291 Skylarks and 689 Hollywoods had been built. After the nominal changeover to 1941 models, 28 more Skylarks were made before Hupp shut down forever. Hollywoods continued to come off the line; 1,173 of them, before Graham converted to war production in September. About 2,200 Hupp Skylarks and Graham Hollywoods were created from the converted Cord dies, far fewer than the number of 810 and 812 Cords that had preceded them.

The Cord dies were tired. Now they could rest.

19
ACD REDUX

Dallas Winslow's new Auburn-Cord-Duesenberg Company sent a mailer to every known owner of an Auburn, Cord or Duesenberg. The response in orders for parts and service was gratifying. A portion of what had been the experimental garage of the Auburn Automobile Company's headquarters building was set aside as a service area. The extensive parts bins, for all of the cars for which the A-C-D Company supplied parts, still filled the once-beautiful art deco showroom.

The company regularly reminded its owner/customer base of the availability of its services. There were still thousands of Auburns, Grahams and Hupmobiles on the road, and hundreds of Cords as well. Through the World War II years, when no new cars were available, a reliable 5- to 10-year-old Auburn was very likely to be in use as everyday transportation.

Many Cords, on the other hand, slept through the war in garages and barns and corn cribs. The likely reason was the Cord's fusible

Leftover sedan bodies, brought back from Connersville, languish behind the Auburn-Cord-Duesenberg Company. The shell in the background was an 810 show car. (Auburn-Cord-Duesenberg Museum)

link—its transmission. By 1939, with no dealer services available and few mechanics versed in the Cord's unique maintenance requirements, abused Cord gearboxes were succumbing to stripped gears and broken thrust washers. Again, the Cords were rescued by their enduring beauty. Too sick to run and too beautiful to junk, many of the cars avoided further destructive wear by being stored until the owner could get around to repairing them.

After the war, business actually picked up in the A-C-D Company's service department. Returning servicemen remembered the old Cord in the barn. Many heroes who did not return left family who wanted to see the young man's old car on the road again. A renewed murmur of interest began in the fixing up of these handsome old cars.

The A-C-D Company was finding little use for the Duesenberg assets it had purchased in 1938. In 1946, it sold its entire stock of Duesenberg parts to Marshall Merkes. Merkes also received the available tooling for making Duesenberg Model J parts, surviving drawings, and the Duesenberg trademark. (His Imperial Manufacturing Company supplied parts and information to Duesenberg owners until 1985. In that year Randy Ema of Orange, California, purchased the assets of Imperial Manufacturing Company, in order to continue to provide parts and service for Duesenbergs.)

Elmer C. Bartels was A-C-D's Sales Manager, and sent out thousands of notices to Cord owners each year. A letter to the A-C-D Company about the availability of a replacement part received an unfailingly courteous response signed "E.C. Bartels." Magazines took notice of the "Auburn Blood Bank" for Cords, and interest grew even more in the cars and in their unique hospital.

The man behind the parts, and behind Mr. Bartels, was Justin Girardot. He monitored the stocks of available new parts, and made the decisions regarding depleted items that might be remanufactured. (When pressures for scrapping parts became intolerable during World War II, it was he who made the decision on what to scrap, too.)

Many of the orphan makes for which A-C-D provided parts could be serviced by a local mechanic, and the company did a big business in mail-order parts. But the Cord's unique mechanicals often required the hands-on talents of someone who understood them. And so, an early trickle of Cords to the A-C-D Company

---

BRING YOUR CORD TO THE FACTORY

Enjoy The Profound Satisfaction It Was Designed To Give By Having It Serviced By Specialists

Owners from every state in the Union are driving their Cords in for inspection and service by our experts. We want to experience the pleasure of receiving a visit from you because we want to meet you personally and then to thoroughly inspect your Cord and give you an honest report on its condition.

Our thoroughly skilled, factory-trained mechanics have had years of experience in repairing Cord cars and know the individual features and peculiar characteristics of each Model and each, by reason of long experience knows—The Most Practical Method—The Quickest Way to Make Repairs of Any Nature.

Every piece of equipment and every special or standard tool that may be of the slightest help has been provided to the skilled mechanics to do the work. Lost motion and waste of time are entirely eliminated. As a result, every repair job is accomplished smoothly in the quickest time and with the least expense to you.

We have a complete stock of genuine Cord parts at new low prices. Never is it necessary for our mechanics to stop work on a job and wait for a part—never does the occasion arise when an old part must be doctored or tinkered into usable condition at a much greater cost and with less value.

All Material and Workmanship are Guaranteed

BRING YOUR CAR TO THE FACTORY

For your own benefit, we urge you to use our complete facilities. You will be amply repaid in the added satisfaction—and savings.

AUBURN CORD DUESENBERG CO
AUBURN, INDIANA

A Dallas Winslow flier.

Posed in the showroom, this Cord is advertising sealed beam conversion kits and add-on windwings. (Ron Irwin)

service shop in Auburn had become a substantial stream by the early 1950s.

A major source of the A-C-D Company's income became the servicing and rebuilding of Model 810 and 812 Cords. The company even began to manufacture some new parts to replace those which so many Cords seem to need. Most popular were transmission gears and universal joints.

A bizarre scenario now unfolded. Jirardot and Bartels were counter men, not engineers. Neither they nor Dallas Winslow were aware that the parts that caused the most trouble in older Cords required particularly careful quality control, and that the specifications for ordering them would benefit from some engineering sophistication. In the absence of such information, universal joints supplied by Bendix to A-C-D were obtained at a lower price by the omission of a critical regrinding step. Gears purchased from Warner Gear's plant right behind the Auburn building were made to the cheapest production gear standards. Good enough for the overdesigned transmissions of Fords and Chevrolet perhaps, but a disaster waiting to happen when installed in the marginal Cord gearbox.

By this time many Cord owners had painfully learned service techniques that the Auburn Automobile Company had never had the time to discover. But the A-C-D Company restored by the book.

Suggestions made by owners for useful modifications were politely turned away. The Cords turned out by their shop, with the best of intentions, were often for sale soon by the disgruntled, poorer owner.

The new A-C-D Club was able to harness the substantial engineering knowledge of its members. They approached Bartels and Girardot with offers of plans and specifications for improved Cord gears and other parts, and sources to fabricate them. The A-C-D Company had recently invested large sums in similar parts, and couldn't understand why theirs weren't good enough. Faced with this, earnest club members began not-for-profit projects to create new parts, properly engineered, for the benefit of their members and other Cord owners. Dallas Winslow, incensed at what he saw as a lack of appreciation, briefly considered junking his stock of new Cord parts. Quiet diplomacy eased the strains.

Auburns and Cords were everyday driving cars in those post-war years. With parts supplied by the A-C-D Company, most local mechanics could still do a creditable job of repairing Auburns. Cords, on the other hand, required unique knowledge and skills. Repair shops specializing in Cords flourished for a period in several parts of the country.

The Bronx in New York City was then a pleasant mix of middle-class apartment buildings and private homes. Marty's Cord Garage on Tiebout Avenue serviced most of the running Cords in northern New York City and southern Westchester County. Liberty Garage in Hillside, New Jersey, was a part-time operation. Its proprietor butchered meat by day. Some wags said that he performed the same service for Cords on evenings and weekends. In San Francisco, Leo Bertelsen repaired Cords and other exotics. *Popular Science* magazine did a story on him in 1949.

Some owners chose not to deal with the cantankerous front drive system at all, and Cords were hacked into rear drive parodies with depressing regularity. California was now deep into the Kustom Kar kraze that it had loosed upon the world, and many a young customizer did his best to improve on Gordon Buehrig's art.

When the Auburn Automobile Company transferred its administration to Connersville in 1938, most of the drawings of Auburn and Cord parts and cars were left behind in Auburn. Some of these dated back to the early part of the century. When spaces were being cleaned out in preparation for occupancy by Warner Gear and Dallas Winslow, some furnishings and artifacts were

---

DRIVING UNIT ASSEMBLY
(Transmission and Differential)

Driving unit trouble usually gives timely warning of its appearance in the form of grinding or growling noises, locking or sticking of gears, inability of gears to remain in mesh or difficulty in shifting from one gear to another.

When any of these symptoms prevail something should be done at once. Rather than to attempt to repair the worn unit it will in the majority of cases be less expensive to install a complete - guaranteed unit that has been properly assembled and tested at the factory.

Driving unit assembly **$97.50**

*Exchange price—To this price add $25.00—old unit to be returned for refund of this amount.

IGNITION DISTRIBUTOR

Satisfactory performance particularly at high speed depends a great deal on perfect timing. You don't get perfect timing from a worn ignition distributor. Of course a worn distributor can be repaired but there is no comparison between one that has been repaired and a guaranteed new one. For the cost of repairs you can install a new one.

Ignition Distributor Assembled
complete *$4.85

*Exchange price. To this price add $2.00—old distributor to be returned for refund of this amount.

ELECTRIC CLOCK

There is no need to be without the convenience of an electric clock (an accurate time piece) in your Cord. If the clock in your car does not run or keep time as it should you will welcome this opportunity to replace it with a guaranteed new one.

Exchange Price *$4.85

*To this price add $2.00—old clock to be returned for refund of this amount.

IMPROVED WHEELS
For Greater Safety

The wheels, particularly the front, are subjected to great stress, especially at the centers where they are bolted to the hubs. This is because of the tremendous power exerted through the front drive.

We now offer an improved extra heavy gauge steel wheel, 20% heavier to be exact, than the original equipment.

Allow us to suggest that you examine the wheels on your car. Should you find the slightest signs of cracks around the hub bolt holes you'll want to effect replacements at once.

Heavy Gauge Wheel .... Price each $8.75

Auburn-Cord-Duesenberg Company prices in the 1950s.

Cords in the A-C-D Company's repair shop. All the cars show at least minor modifications. Notice the outside exhaust pipes that rise with the hood! (Auburn-Cord-Duesenberg Museum)

auctioned off. Old documents, including full-size body drafts, were held in such low esteem that the going price was a penny apiece. A local signpainter bought hundreds of them for use as dropcloths and for making stencils. Randy Ema rescued the last of these years later just before the sign company discarded them. Most of the remaining drawings had been thrown away by the A-C-D Company in the 1940s.

Dallas Winslow was a remarkable entrepreneur, and had long recognized the value of devoted employees. He was providing his workers with health benefits and insurance long before these became the norm. Beginning in 1950, he gave every employee of his companies a new Ford V-8. (He later switched to Chevrolets.) Winslow retained title, but the employee could use the car as his/her own. Each year the cars were replaced by new ones.

Winslow now had a growing stock of used cars to dispose of. They were reconditioned in the service department of the Auburn-Cord-Duesenberg Company, and started to slow the work on restoring Auburns and Cords. By 1959, the used Fords were starting to clog the former factory showroom. It was time for the next step in the Cord saga.

Glenn Pray started teaching shop in a high school in Tulsa, Oklahoma, in 1952. He found his first Cord by posting pictures of what he wanted on the school bulletin board; students went looking, and found one for him. By 1956 Glenn owned 13 Cords, and attended a car show in Indianapolis where one of his restored Cords won first prize. It rained the second day of the show, so he drove up to visit the Auburn-Cord-Duesenberg Company in Auburn. That was the start of his determination to someday own the company.

Dallas Winslow's offices were then in Detroit. Pray called him in 1960 and asked if the company might be for sale. After some guarded reticence, Winslow agreed to meet Pray in Auburn. Winslow was no longer in the best of health, and had not been pleased in recent years by what he saw as the ingratitude of Cord owners toward his investment in their cars' future. After some haggling, Pray emerged with an option to buy the 700,000 pounds of parts that filled the old Auburn headquarters. It wasn't easy for a penurious schoolteacher to raise the $35,000 needed for the down payment, but by ingenuity, good fortune and just plain begging Pray was able to meet Winslow's terms. Pray gives special credit to his

Top: Pray's Cords in production in 1964.

Bottom: Glenn Pray

Lightly used Chevrolets and Fords fill the Auburn showroom in 1957. (Auburn-Cord-Duesenberg Museum)

partner and friend, Wayne McKinley, who helped with that initial financing.

Pray and friends moved the parts from Auburn to Broken Arrow, Oklahoma, in used trucks that he bought then resold when the job was done. About 30% of the parts were for the 810 and 812 Cords. The rest were mostly for Auburns. Among the tens of thousands of pieces were 100 Cord instrument panel face plates, nearly 1,000 universal joints, and 300 heavy-duty wheels.

Pray juggled creditors, sold parts and worked toward the goal for which he had bought the company in the first place. Glenn Pray intended nothing less than the production of a modern car with the classic appearance of the original Cord. In 1963 he shared his ideas with Gordon Buehrig, who offered to sculpt a quarter-scale model of a smaller 2-passenger version of the original 810. Gordon actually began this work, but his employer, Ford Motor Company, considered Pray a competitor. Glenn considered that a great compliment.

How Glenn Pray brought his 8/10 scale new Cord to market is a story in itself. That he did it at all is a tribute to his dogged talent, and an even further tribute to the timeless Cord design, which appealed to the finer instincts of car-lovers in 1963 as it had in 1936. With rare foresight, Glenn set up the firm that was to produce the new Cord as

They look like junk, but in the hands of Leo Bertelsen these cars will get a new start in life. Shown here are four Cords, a Duesenberg (at the left in back row), and an old Studebaker.

Good Cars Never Die

That's the working philosophy of Leo Bertelsen, former race mechanic who rebuilds many fine cars of the past.

By Murrelle Mahoney

SOME of the finest cars in the West are overhauled in an unpretentious one-man garage in San Francisco. The secret is the man who runs it, Leo Bertelsen. A mechanic's mechanic, he has an unusual knack for rebuilding high-performance autos.

Bertelsen's start in the footsteps of Gus Wilson was inauspicious. At the age of 12, he

Cords are the specialty of Bertelsen's repair shop. Although all of these cars are now at least a dozen years old, roadworthy examples still command a stiff price. Here, Bertelsen checks over the V-8 engine of one job (left) and the front-drive transmission of another.

> ### The Tucker Raids
>
> A substantial percentage of spare transmissions were lost to Cord owners when Preston Tucker embarked on his ill-fated assault on Detroit in 1947. To buy time for Tucker engineers to design and tool up for an automatic transmission, chief draftsman Fred Loetterle suggested the use of a Cord transmission mated to the Tucker's rear engine. Preston Tucker Junior and others scoured the country and located twenty-two Cord transmissions. The first twelve Tuckers had reworked Cord transmissions; later ones used a newly-manufactured gearbox. It had four synchronized forward speeds, and was a bit longer. But it was still shifted by the Cord's electric-vacuum mechanism, and still based on Harry Weaver's thirteen-year-old design.

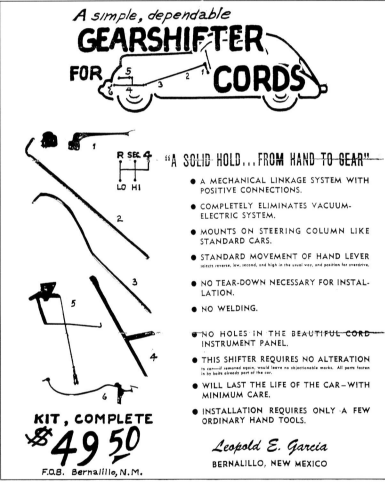

a corporation separate from his Auburn-Cord-Duesenberg Company. He licensed the use of the Cord name to it, with the proviso that the right to the name revert to him if the car went out of production.

The project didn't succeed, and was sold at auction in 1966. The successor company briefly produced a series of cars with the Cord name on them. The horribly distorted body lines were more travesty than tribute.

While engaged in the new Cord project, and in a later one creating a fiberglass-bodied car in the image of the 1935 Auburn Speedster, Pray sold original Cord and Auburn parts. Over the years the 700,000 pounds of original stock was seriously depleted. Some parts were newly manufactured, as they were needed. The title of "the factory" was bestowed by enthusiasts on Glenn Pray's Auburn-Cord-Duesenberg Company. It became a priceless source of original Cord parts, and a reference source for even more priceless accurate information.

Top: The sign outside the former pickle factory that housed Glenn Pray's Auburn-Cord-Duesenberg Company.

Bottom: One of several aftermarket attempts to replace the Bendix electric-vacuum gearshift was this mechanical shift linkage sold in the 1950s.

20

THE ENTHUSIASTS

The National Used Car Market Report is better known as the "Blue Book." In 1940 it told a depressing tale of a unique orphan car that few wanted. A 1936 Cord Westchester sedan shows an average sale price of $210. A supercharged 1937 convertible coupe would bring $300. Those represent losses of about 90% of the car's value in four years. (For comparison, a 1936 Oldsmobile touring sedan that sold for only $950 in 1936 was still worth $370.)

Owners of old Packards and Cadillacs love their cars. But few of them can describe, in the most intimate detail, the first time they saw one. Nearly every Cord owner can tell you his age when he saw a Cord for the first time, its color, body style, the weather and the circumstances. The impression is that strong, and it hardly ever goes away.

This powerful passion for the car, and the special mechanical needs mandated by its unique construction, make Cord owners a breed apart. So while the parts made available by Dallas Winslow's

The cars of the Maryland Auburn-Cord Club, 1941. (Ed Kairis)

223

SELL—Late '51 Nash-Healey; maroon, whitewalls, roll-up windows, 2400 miles, like new. Delivered price $4300, will sell for $3750. Need closed car. J. D. Powell, 3100 West Cary St., Richmond, Va. Phone 6-4646 or after 7 p.m., 88-1225.

WANTED—Correspondence with anyone interested in forming a club for the restoration and preservation of Auburn, Cord and Duesenberg automobiles. H. Denhard, R.D. 2, Greenville, N.Y.

SELL—'31 Victoria Rolls-Royce five-passenger conv.; good engine, tires, needs paint, top upholstery. Terrific car to customize, $950. Photos, complete info on request. E. Woelffer, 21 West Boulder St., Colorado Springs, Colo.

SELL—'33 SJ Duesenberg phaeton; new tires, battery, valves just ground, new upholstery, runs like new. $1750. C. Grosscup, 1121 Van Buren NE, Auburn, Ind.

WANTED—'32 Oldsmobile shop manual six and eight cyl., also hub caps and speedometer and cable for Oldsmobile Eight. Must be in good condition. O. Elster, Box 273, Sausalito, Calif.

WANTED—Help in refinishing '29 Packard, model 640. Would like to correspond with other owners. H. Gustafson, Wooster St., Ridgefield, Conn.

WANTED—Body, hood, radiator, shell and fenders for '15 four-cyl. Saxon. R. Young, Stocksville, N.C.

SELL—'31 Auburn eight sedan; dual manifold and new electric fuel pump, takes Ford Six or Chevrolet carbs. Car is restorable, needs ignition work. G. Mitchell, 3880 Vorhies Rd., Ann Arbor, Mich.

WANTED—Columbia rear end complete with all accessories for '41 Ford in good running order. From LA or nearby. D. O'Regan, 1322 E. Acacia

The ad that started the Auburn-Cord-Duesenberg Club.

company kept the surviving Cords staggering along, Cord owners badly needed a way of communicating, commiserating and sharing. And like-minded folks have a way of finding each other.

In the Chicago area as early as 1939 the Cord Owners Club (Illinois) was established. Car collector D. Cameron Peck was president. Dues were $5 per year, a sizable sum then. Only the actual owner of a Cord was permitted to join. The club's purpose was . . ."the banding together of all owners of Cord cars in the State of Illinois for the protection and enhancement of their mutual interest as owners of one of the world's finest and most interesting automobiles."

If a member sold his Cord and did not intend to purchase another he was obliged to resign from the club within thirty days of the sale. The club published a little "Constitution" that's so stuffy it must have been tongue-in-cheek.

In the Baltimore area before World War II, Ed Kairis, Millard Wilkerson and others banded together as the informal Auburn-Cord Club of Baltimore. Club members owned two Cords and six Auburns. The club disbanded in 1943 as members were drafted into the armed services.

No passenger cars were manufactured anywhere in the world between 1942 and early 1945. During those years people drove anything that could be made to run. Gasoline rationing reduced travel miles.

After World War II, American car production resumed. True, most 1946 and 1947 models were warmed-over versions of 1941 and 1942 cars, but the public was so car-hungry that it bought whatever the manufacturers could produce. By 1949 nearly every American car was of new post-war design. The cars of the thirties were becoming relics of bygone years, and were driven by those who could not afford new cars, and by young people looking for their first car.

During these years surviving Cords were still bought and sold for small sums. Sedans brought around $300. A supercharger added about $50 to the price. The open cars commanded comparatively handsome figures of up to $1,000.

(l-r) Augie Duesenberg (holding hood), Ab Jenkins in Mormon Meteor, and Harvey Firestone with Firestone executive, in front of the Firestone building.

FOR THOSE WHO HAVE NEVER RELISHED THE COMMONPLACE

Auburn-Cord-Duesenberg Club Newsletters, forty years apart.

An early A-C-D Club reunion in Auburn.

The Cord Car Club that Theodore J. Ciuba started in Wichita Falls, Texas, in 1951 was a very informal organization. Ciuba told me that he was the president because he owned the only complete set of Cord service bulletins that his club members knew about! Members would write to him for the factory "gospel." He also served as a clearinghouse for the buying and selling of parts.

In 1952 in upstate New York Harry Denhard owned an Auburn Speedster. He had previously owned a Duesenberg Model J, and a Cord 812. Harry thought that owners and fans might want to help each other preserve and maintain these magnificent relics of America's automotive past. So he placed an ad in *Motor Trend* magazine:

"WANTED—Correspondence with anyone interested in forming a club for the restoration and preservation of Auburn, Cord and Duesenberg automobiles. H. Denhard, R.D. 2, Greenville, NY."

Thirty-five people responded to Harry's ad, to his pleasant surprise. He named the group the Auburn Cord Duesenberg Club, and added the motto "For Those Who Have Never Relished the Commonplace." Beginning in September 1952, Harry turned out a small mimeographed monthly newsletter including letters from members, ads selling or seeking cars and parts, and the sharing of technical information. Harry supported the Club himself for the first several months of its existence.

He instituted $1.00 annual dues in January 1953. By September, the Club had grown to over two hundred members. Committees were formed. Technical and historical information flowed in. Over half of the cars owned by club members were 810 and 812 Cords, more than all the Auburns, Duesenbergs and L-29 Cords put together.

Clearly, Harry had tapped into a vein of enthusiasm, and it was getting beyond what he could handle by himself. By 1954 costs of publication and mailing had risen, and Harry suggested an increase in dues to $3 per year. Many members were not pleased, and in February 1954 Harry gave it up.

Volunteers from around the country stepped forward to pick up the burden. Their efforts assured the continuing exchange of information and parts that would keep the unique machines alive. Bob Fabris of Maryland took on the task of Membership Secretary and Treasurer. Al Kimber and Reg Taverner of California mimeographed, stapled and mailed the monthly *Newsletter*. Hours of gratis technical support to Cord owners were provided by Taverner and by Jim Howell of Chicago. Jim also engaged in voluminous correspondence, dispensing invaluable free advice. Bob Fabris, Bob Graham of Kansas and Jim Hoe of Connecticut did the same for L-29s, Auburns and Duesenbergs.

The club's publication became the glue that held it together. The new *Newsletter* editor, beginning in April 1954, was Bob McEwan of New Jersey. For the next three years, Bob edited the *Newsletter*, encouraged new members (including me), managed meets, and

Top: The low point: the former Auburn showroom in use as a machine shop by the A-C-D Company. (Auburn-Cord-Duesenberg Museum)

Bottom: Gordon Buehrig at an A-C-D Club reunion

Days of glory again: the restored showroom of the Auburn-Cord-Duesenberg Museum.

generally willed the infant club to grow through the force of his own personality.

The first meets of the Auburn Cord Duesenberg Club took place in the East. Gordon Buehrig attended them; he was somewhat bewildered by the attention of Cord fans who hung on his every word.

In 1956, A-C-D Club board member Joe Knapp conceived the idea of a gathering in the city where it had all started. Roy Faulkner then lived in Fort Wayne, just down the road from Auburn. Knapp called on him for support. Members and guests were invited from all over the country. Knapp and volunteer Del Johnson of the town of Auburn viewed the fall weekend event as a reunion, because the men who had built the cars and had run the company were to attend as well.

Gordon Buehrig was there. So were Harold Ames, Alex Tremulis, Herb Snow and Cornelius Van Ranst. Two expected guests passed away only months before the meet: Ab Jenkins, and Roy Faulkner himself. And E.L. Cord? He was invited, but present in spirit only!

The Auburn-Cord-Duesenberg Company hosted the Auburn Cord Duesenberg Club. Club members thought they had died and gone to heaven, gazing at rows upon rows of bins holding nearly every component part of their beloved Cords and Auburns. And prices, to boot, were comparable to those of similar parts for new cars.

That first reunion helped the town of Auburn remember its past, and take pride in it. And from it grew an event for which tens of thousands gather each Labor Day weekend. By 1990 the annual Auburn-Cord-Duesenberg Festival had become second only to the Indianapolis 500 race as an Indiana tourist attraction.

A centerpiece of the annual event is the former showroom and administration building of the Auburn Automobile Company, where our story began so many years earlier. Used as a warehouse by Dallas Winslow, the building was later subdivided and occupied by several small manufacturers. It was damaged in a 1974 fire. Rescued by Auburn residents who turned it into the Auburn-Cord-Duesenberg Museum, the building has been restored to its original grandeur. As in the 1930s, Cords, Auburns and Duesenbergs grace the terrazzo floor of its showroom. A remarkable library is a central resource for historical research on the cars of Indiana in general and of E.L. Cord in particular.

Gordon Buehrig once described the members of the Auburn Cord Duesenberg Club as "... crazy, lovable people ... who absolutely refused to let these cars die."

It's because of these enthusiasts that this book is about the history of an automobile and not about a museum piece. Because of their special affection for this Hoosier masterpiece, Model 810 and 812 Cords still motor happily along the highways of the world.

AFTERWORD

The question is often asked: What impact, if any, did the spectacular Cord Models 810 and 812 have on the automobiles that followed them? The answer is, a great deal and very little.

Auburn's stubborn efforts to perfect a front wheel drive system were certainly prescient. At the time this book is being written, 95% of the cars being built in the world pull instead of push. But Citroen mass-produced front drive cars before the 810 Cord, and for decades afterward. And the breakthrough that made front wheel drive really practical came many years later, in the transverse-engined cars pioneered for production by Britain's Alec Issigonis.

Herb Snow and his engineers had indeed created a uniquely engineered new automobile with an unusual and forward-looking drive train. Under the most extreme pressures of time and budget, they got it at least 85% right. They were talented men, and the Cord does them proud. But the world didn't beat a path to front wheel drive because of Auburn's pioneering efforts. And, truth be told, little else about the Cord's mechanical design was unique.

Bits and pieces of the influence of the Cord body and interior styling did work their way into other cars: the low outline and lack of running boards; the "step-down" entry; the concealed gas cap and door hinges, and the horn ring; the fastback body; years later, concealed headlights. Looked at from these perspectives, the Cord 810 and 812 were interesting, but at best historical footnotes.

So it wasn't the mechanical features that made the Cord immortal, and it wasn't the design details. Why, then, did *Automobile Quarterly* once state that they received more letters asking for articles about the Cord 810 than any other car?

It was the *look* of the thing! That glorious, shimmering visual unity. That play of line and form so satisfying to the senses. That urge to seek new angles from which to examine and enjoy it. A viewer's tribute to the designer may be apocryphal, but is nevertheless appropriate: "It didn't look like a car. Somehow it looked like a beautiful thing that had been born and just grew up on the highway!"

Add to that the remarkable *synergy* of body and chassis. The Cord's unusual styling made it stand out from the crowd. The mechanical muscles and sinews that powered this breathtaking creation just had to be something out of the ordinary, and the Cord did not disappoint. Snow's engineering and Buehrig's art rose together to heights that neither might have attained alone.

Glenn Pray's creation of a new Cord in 1964 was high tribute to the original. Another tribute is less well known. The first front wheel drive automobile built in the United States after the demise of the Cord was the Oldsmobile Toronado of 1966. Cord aficionados have noted sly resemblances in the later car. Concealed headlights, perforated wheels, louvered grille, the fastback rear and a minimum of trim. Somehow, the same feel was there.

The basis for the Toronado's body design was a full size airbrush rendering done in the Olds design studio by its assistant chief designer, David R. North. The designers called it "The Flame Red Car." North later told Michael Lamm, editor of *Special Interest Autos,* that he remembered ". . . having photographs on our desk of the Cord and trying to get a little of that romance into it." Three decades earlier, Gordon Buehrig had lost a design contest at GM with an idea that he later developed into the Cord 810. Thirty years divided the Red Clay Model from the Flame Red Car. General Motors had finally gotten the message.

In 1951 New York's prestigious Museum of Modern Art staged an exhibit called "Eight Automobiles." A Cord sedan was one of them. *Fortune* magazine polled industrial designers the world over in 1959 for their selection of the 100 best mass-produced objects ever. The Cord 810 was number 14. A 1987 exhibit of modern design at the Brooklyn Museum was called "The Machine Age." Its automotive icon was a Cord 810 Westchester sedan. In 1982 *Motor Trend* magazine asked the top stylists of all the domestic auto makers and several foreign ones for their choices of the ten most beautiful cars ever built—"the best examples of clean, pure and beauteous design. . . which turn you on personally, excite your aesthetic libido, and generate a desire to possess. . ." Their overwhelming favorite was the 810/812 Cord.

Gordon Buehrig and Herb Snow and the men and women of Auburn were in the car business, a pursuit that's almost a synonym for planned obsolescence. Somehow, their remarkable creation remained virtually undated. The Cord's time was yesterday and today and tomorrow. The poet was right.

A thing of beauty is a joy forever:
Its loveliness increases; it will never
Pass into nothingness

— "Endymion," John Keats.

BIBLIOGRAPHY

Books

The Classic Cord, Dan R. Post Publications, 1952. Second Edition, 1954.

This was the first book dealing with the Model 810 and 812 Cord. It included material on the Model L-29, but two-thirds of the pages were devoted to the later Cord. Six pages of text were supplemented by reprints of Cord ads, brochures and information for salespeople. Information given was the best that was known at the time. Many car fans of the 1950s were converted to Cord enthusiasts by this powerful volume. Hardcover and softcover editions were available.

The Cord Front Drive, Roger Huntington, Floyd Clymer Publications, 1957.

This softcover volume was written with Huntington's trademark flair, and is entertaining reading. The L-29 is discussed, and much factory and trade material reprinted. Cord owners of the 1950s shared their opinions of their cars. Huntington specialized in slide-rule analyses of historical vehicles, and compares both models of Cord with contemporary vehicles and with cars of their own era.

Cord Model 810 & 812 Owner's Companion, Post-Era Books, 1975.

Author-publisher Dan Post had earlier expanded the L-29 section *of The Classic Cord* into a book of its own. While collecting material for a never-completed similar volume on the Model 810 and 812, he issued this useful compendium of factory literature, owner's testimonials, and maintenance resources.

Rolling Sculpture, Gordon M. Buehrig with William S. Jackson, Haessner Publishing Company, 1975.

The stylist of the Cord discusses his long and fruitful career in the automotive industry. Forty pages are devoted to Gordon's personal recollections of his role in the Cord 810 project. With this memoir, Buehrig shared the first detailed insights into the workings of his automotive styling studios.

Auburn Cord Duesenberg, Don Butler, Crestline/Motorbooks International, 1991.

This book contains reproduced photos of cars produced between 1901 and 1980 by enterprises using the Auburn, Cord, or Duesenberg names. Accompanying text includes two pages on the Cord Models 810 and 812. The author passed away during the research stage of this project. The book was completed and published by others, who did not have the benefit of Butler's command of automotive history.

Errett Lobban Cord: The Man, His Motorcars, Griffith Borgeson, Automobile Quarterly Publications, 197?.

Borgeson's masterwork describes the life of E.L. Cord and his myriad entrepreneurial activities. Years of research uncovered the details of the deals and corporate complexities that marked Cord's meteoric career. Fine photographic coverage is given to the cars as well, including the Models 810 and 812. This massive leather-bound volume was sold in a limited edition, each copy numbered and signed by the author.

Periodicals

Literally hundreds of magazine articles have been published about the Model 810 and 812 Cords and allied subjects. Many are accompanied by photographs, some historical, some of restored Cords. Accuracy of the text and authenticity of the photographs is—how to put this delicately?—inconsistent.

The articles listed below are a sampling of my personal favorites.

"Prototype of the Future," George Finneran, *Motor Trend,* July 1950.

"Auburn Blood Bank," Roger Huntington, *Motorsport,* October 1951.

"Pull Instead of Push," Ken Purdy, *True,* January 1952. (Later reprinted in his anthology *The Kings of The Road.*)

"Cord-Auburn Health Center," *Car Life,* February 1954.

"Salon: Cord 812," *Road and Track,* June 1957.

"Cord Off The Cuff," Glenn Pray, *Automobile Quarterly,* Volume 4 Number 2.

"Cord 810," Wim Van Der Graf, *Car Life,* March 1965.

"Did The Cord Really Contribute?" Roger Huntington, *Car Classics,* August 1975.

"E.L.: His Cord and His Empire," Beverly Rae Kimes, *Automobile Quarterly*, Volume 18, Number 2.

"Toro and Cord," Michael Lamm, *Special Interest Autos # 35,* July-August 1976.

"Cords In Disguise," Jeff Godshall, *Special Interest Autos # 65,* October 1981 and #66, December 1981.

"Greatness Gone Awry," Karl S. Zahm, *Collectible Automobile,* December 1986.

"1936 Cord 810 Westchester," Arch Brown, *Classic Auto Restorer,* June 1993.

Organizations

Two organizations are unique sources of Cord-related information. The true enthusiast should join both.

The Auburn Cord Duesenberg Club. The club's *Newsletter* has been a treasure trove of published personal memories, photographs, speculation, and rare data since 1952. Club members receive the publication ten times each year, and may purchase available back issues.

The Auburn-Cord-Duesenberg Museum. The museum's quarterly publication, frequently offers rare photographs gleaned from the museum's wonderful archives. The museum's library is the most important repository of primary reference material for those researching the history of cars of Indiana, and particularly the three cars for which the museum is named.

APPENDIX I

Production Statistics

No certain figure can be given of the number of Cords produced. Approaching an estimate from three different perspectives permits a reasonably accurate guess.

Chart A is based on the highest Central Manufacturing Company number known, for each body type. Central began the numbering for each type with 101. The showcars are included in these figures, since they were included in the numbering sequence. Based on this calculation, 2,972 Cords 810 and 812 were built.

The showcar figures are not included in Charts B and C. Chart B is based on numbers given to Ron Irwin by Don Butler, who had been the Car Distribution Manager for Auburn in 1936 and 1937. Chart C is based on the difference between the highest and lowest known serial numbers for each area of production.

The exact number of showcars that were eventually built is not definitely established. (It should be noted that the great majority of showcars were fastback Westchester and Beverly sedans.) The highest estimate for showcars built is 92. This is Don Butler's figure, and was corroborated by what Slim Davidson also told Cord scholar Paul Bryant.

Adding 92 showcars to chart B's total gives 2,992. Added to Chart C's total, the result is 2,999. It seems reasonable to state, therefore, that between 2,972 and 2,999 Cords were manufactured in 1936 and 1937.

2,320 Cords were registered in the 48 states in 1936 and 1937. About 10% of production was exported, or about 300 cars. Auburn had about 250 dealers in August, 1937. If the average dealer had one or two cars in stock, a reasonable assumption, that could account for another 375 cars. The total of 2,995 is close to the production estimates above.

Cord Model 810 and 812 Production

CHART A

		By Highest Known Body Number	Cars	Percent of Total
C	90	Fastback sedan (Westchester and 810 Beverly)	1474	50%
C	91	Convertible Phaeton Sedan	610	21%
C	92	Convertible Coupe	205	7%
C	94	Early 810 Beverly	5	
C	95	Custom Berline without bustle trunk	2	
C	96	Bustleback sedan (812 Beverly)	401	13%
C	98	Experimental Convertible Coupe	1	
C	99	Coupe	2	
C	100	Convertible Phaeton Sedan, bustle trunk	2	
C	101	Custom Beverly sedan, 6-window	3	
C	102	Convertible Coupe with rumbleseat	1	
C	103	Custom Berline	36	1%
C	105	Custom Beverly	229	8%
C	106	Special Custom Berline, 135" wheelbase	1	
		Total cars	2972	100%

CHART B

From Company Data	unS/C	S/C	Total	Percent
Westchester	1228	81	1309	45%
Beverly	327	184	511	18%
Convertible Phaeton Sedan	404	196	600	21%
Convertible Coupe	130	64	194	7%
Coupe	2	1	3	
Custom Beverly	94	141	235	8%
Custom Berline	27	21	48	2%
Total Production cars	2212	688	2900	100%

Source: Don Butler, Car Distribution Manager, Auburn Automobile Company

CHART C

By Highest Known Serial Number	Lowest	Highest	Total	Percent
1936 standard wheelbase	1101	2729	1629	56%
1937 standard wheelbase	1526	2522	997	34%
1937 Custom series	10001	10281	281	10%
Total Cars			2907	100%

APPENDIX II

Reports on the Prototype's Test Run

On August 5, the day E-306 #2 headed back for Auburn, Herb Snow prepared a memo listing the complaints he had heard by telephone from George Kublin in Los Angeles. This document amounts to a synopsis of Kublin's yet-to-come final report on the trip, with recommendations for action. (It's fortunate that we know which car was on its way back from Los Angeles; we would not recognize it from the heading.)

COMPLAINTS ON FIRST E-294 CAR

BRAKES

These brakes are getting hot and scoring. Layout should be made of the Bendix brakes for 11 x 2 1/4" size, using centrifuse drums. Two sets of these should be ordered and rushed through as quickly as possible.

FUEL PUMPS

They have experienced trouble with vapor-lock, which is due to either the fuel pump getting too hot or the carburetor bowl. Should consider using asbestos gasket between the carburetor and head.

THROTTLE AND CHOKE CONTROLS

Print should be changed to use rods in place of bowden wire for both of these controls, and some friction device added which will hold lever in any position.

BRAKE AND CLUTCH PEDALS

There is too much frictional resistance on the hook-up for these pedals, both brake and clutch. Effort should be made to free up all of the parts in this hook-up.

WIRE HARNESS

A complete harness should be worked out at the first opportunity and a sample braided harness made up to these specifications. In making this layout, be sure that it is properly clipped so that none of the parts hang out in the open.

EXHAUST PIPE

Exhaust pipe should be clamped to the engine at a point near the junction of the two front pipes, so that there will be no motion between this pipe and the exhaust manifold. Flexibility should be between this point and the pipe attachment to the frame.

INSTRUMENTS

The hands on the instruments should be wider and the color changed, as it is very difficult to see these at night. Mr. Thomas will take this matter up with Mr. Maier of Stewart Warner and have one set of these instruments changed over.

UNIVERSAL JOINTS

The Weiss joints are still noisy at the extreme turns, also experienced difficulty with oil leaking out. Order for Rzeppa joint should be confirmed and sample rushed as much as possible.

BATTERY

We should use larger capacity battery on all of these cars. Mr. Thomas will see what the cost would be to get a lower battery. Mr. Allen will see if it is possible to lower the front seat provided a lower battery can be obtained.

BRAKE HOSE

Wagner Service Station in Los Angeles advised that the hose which we used on this car is a type which has been obsolete for the past two years.

MUFFLER

The muffler on the car is very noisy and tests should be made as quickly as the new body is mounted on the experimental chassis to get a satisfactory muffler.

ENGINE

This engine uses too much oil. I will write to Mr. Baster of Lycoming, insisting on the use of four rings on all of these pistons.

HEADLAMP

Have had considerable trouble with the headlamps on first car rattling. Will rush through the samples of the two new mountings as quickly as possible and have one set of each installed in fenders for test.

FRONT FENDER

An additional brace should be provided at the front end to give more rigidity.

WINDSHIELD

Windshield should be changed for the third body, raising the top of the frame 1" at the center and 1-3/8" at the end.

One set of windshield frames should be returned for reworking and necessary changes in third body made to accommodate same.

REAR WINDOW

Increase the height of the rear window 1" and change length in proportion.

A few days after the return from the West Coast, Thomas sent a memo to Snow and to his fellow travelers. In it he describes the work done to date on the test car.

**WORK DONE ON E-294 CAR #2 UPON RETURN FROM CALIFORNIA.
ODO - 5564**

New water hose was installed on job with 2 laps of asbestos wrapped so heat from exhaust pipe would not strike.

Muffler, which was an Oldberg, was removed, and a new Buffalo muffler, #4586, was installed.

Carburetor adjustment screws were removed and screwdriver slots were cut in them.

Removed radiator core and a new one which is 1" higher was installed. Corners on top tank of radiator were not cut off so these corners were bent down. This had to be done so hood could be closed.

Cylinder heads were removed and carbon cleaned. Top of pistons were checked and found sharp edge had been broken. Also carbon was cleaned from pistons. Valves were removed, found intake valves in good shape but exhaust valves were not seating, so intake and exhaust valves were refaced, then ground in. Carbon on these parts was very heavy for only 5564 miles.

Shock absorbers were removed and a complete new set of shock absorbers, which are inertia type, were installed.

All wheel bearings were checked and greased where necessary.

Removed oil pan unit and replaced with new unit. Checking, found unit was okeh. All trouble was in gauge in dash. Found point in unit shorting against dash. Bent point away from dash, now making gauge work okeh.

All universal joints were removed and inspected by representatives from Bendix Corp. and by Mr. Kublin and Mr. Thomas. Left hand side inner and outer joints were okeh. On right hand side inner joint was okeh, but outer joint had looseness, so a new joint is being installed which was received from Bendix Corp. This joint has the balls pre loaded.

Brakes were checked and readjusted.

Production Electric Type horns were removed and new brackets were made, then horns were installed.

Pedal guides were checked and found fairly free, so no work was done on these guides.

Radio aerial brackets, which were plates, were removed and installed our production running board aerial brackets which are rubber on one end.

Exhaust pipes were removed and a metal ring was installed. This will keep pipe from working out of exhaust manifolds.

Distributor was removed and found points pitted very badly so a complete new set of points were installed. Spark plugs didn't look very good so a complete new set were installed, setting gaps at .027.

Removed gas tank and checking over suction line found it short in length and high from bottom of tank. As close as checking could be made, it showed from 5 to 8 gallons short, or that is it would start sucking air with this much gas in tank. Unit was removed and a new line was installed which should make tank okeh.

All splines were greased with heavy grease and a felt seal was installed on splines. Nuts were counter-bored so felt will fit in bore and when nut is tightened it presses felt into splines. These seals were installed to stop grease from working out of splines, then getting on brake shoes and drums.

Oil and water caps on cowl fillers were replaced with new gaskets. A paper gasket was removed from water filler cap and a new cork gasket was installed. On oil filler cap a shim and a new leather gasket were installed.

Bayonet gauge stick in oil pan was removed and had a piece put in so now oil can be checked very easily.

Screws were put in holes from which sun visors where removed.

Gas Gauge not working. Checking over, found tank unit not working, so a new unit was installed now making okeh.

8-98 style hood bumpers were installed between radiator core and chrome louvers.

Before these were installed louvers would move from side to side when hood was raised and lowered. After these bumpers were installed louvers are solid.

A metal spring was installed on gas tank cover. This spring will keep lid closed but with the light hinge on lid and spring being on side it lets cover cock to one side. Removed exhaust pipe bracket from fender brace. Installed a new bracket on side rail of frame using our rubber mounting.

Put wadding between body pan and top of gas tank. Also made two pieces and put on bottom of gas tank to take out drumming noise and stiffen bottom of tank.

A new battery hot cable was installed. Had to remove terminal end and turn over and by doing so made a neater installation.

Radio was fixed three times then wouldn't work due to when radio set was bolted to dash it put a bad bend in volume control. Spacer was installed between dash and radio set, which now makes radio set work okeh.

Installed sway bar between rear shock arms.

Kublin submitted his comprehensive written report on August 21. By that date many of the problems revealed were already being worked on. Some would be solved before production started, others not until much later. Some would never be completely solved.

Car weight 3,800 pounds, fully equipped, including water, oil and gas. Wheelbase, 125". Engine, V-8, 3-1/2 x 3-3/4—288.6 cu. in. displacement. Total mileage, 5,203. Total gas consumption, 324 gal., or 16.1 m.p.g. Total oil consumption, 19.2 qts., or 267 m.p.g.

TRANSMISSION Gears in all speeds extremely quiet. Continuous trouble experienced with jumping out of second speed under pull at 20-35 m. p. h. This condition did not occur on coast. On return trip, after approximately 4,900 miles, at Dixon, Illinois, in starting up car, and without warning, transmission locked up tight, and car could not be driven either forward or back. Transmission replaced and trip continued. Apparently trouble due to failure of thrust washer, causing gears to jam. See separate report for complete details on condition of transmission.

UNIVERSAL JOINTS Pronounced rattle and knock in joints on extreme turns. Particularly pronounced in right joint. Lubricated right joint twice and left once on trip west. Necessary to disassemble joints to lubricate thoroughly. Inspected joints in L.A., and found excessive play on extreme angle in right assembly. Replaced with new unit. However, this did not eliminate knock on extreme turns. Both joints lubricated and no further lubrication on return trip. Inspection on return showed leakage but still sufficient lubricant in joints. Excessive play in left front wheel caused by wear on joint thrust washer. This condition apparently explains excessive wear on left front tire.

BRAKES Source of continuous trouble on entire trip. Ran extremely hot under normal applications, causing brake fluid to gasify and complete failure. Excessive heat also resulted in grease running out of right rear wheel, causing rear wheel bearing to freeze up. Polished bearing and packed with fibrous grease, and no further trouble experienced. Bearing replaced at L.A. Left rear brake hose failed completely, due to fitting pulling out. Wagner station at L.A. advised that this was old type assembly. Both rear drums badly scored and shoes worn, necessitating replacing at L.A. A heavier brake fluid was used, but even this did not materially help situation. Experience with brakes indicates that size is wholly inadequate for car weight and high speed stops. Very poorly ventilated. Present brake 10 x 2. Sample set of 11 x 2 with centrifuse drums and openings in wheel for ventilation now being installed for test purposes.

SPRINGING Riding qualities in both front and rear seats for city driving, as well as high speeds, unsurpassed ever. Mr. Cord objected to lean-over or roll and excessive tire squeal in negotiating sharp turns, while roll is not particularly noticeable or felt by the occupants of the car, it is readily observed in watching the action of the car on the road. A roll rod has been installed in the rear of the car, and this shows a definite improvement in respect to elimination of tire squeal, as well as lean-over, without materially affecting riding qualities. The trip west was made without the 3/4" blocks, between the rear spring and axle, and with three passengers and a considerable amount of luggage, at no time did the rear springs bottom. However, on the return trip the blocks were installed and on several occasions under extreme conditions the rear end bottomed. It is, therefore, questionable whether the car can be lowered this additional 3/4" at the rear.

STEERING Steering excellent, with no trace of wander under all road conditions and at all speeds. Mr. Cord objected to the heavy pull on the steering when going in and coming out of a curve. Driving the car on a hot day, perspiration from the hands renders the steering wheel very slippery, giving one a feeling of insecurity. This, in my opinion, is serious, and a change in the finish, and perhaps the section of the rim, should be made, to insure a more positive grip.

COOLING SYSTEM Although temperatures as high as 115° above were encountered, no water was added on the trip from Auburn to L.A. The water temperature on the indicator showed an average of 185° under high speed driving in the heat of the day, and the air temperature in the front compartment with the one cowl ventilator open at no time caused any discomfort. The unusual louver treatment probably explains the lower front compartment temperatures, which is a definite improvement over any of our previous models. Mr. Cord objected to the fan noise, particularly in first gear, and by reducing the diameter from 18" to 17" this condition was improved. However, further work should be done in an effort to quiet the fan. On return trip, in driving through the mountains east of Salt Lake, and at an air temperature of 98°, the water temperature showed 200° on the indicator, and in stopping the car the radiator boiled. The smaller diameter fan and a tail wind coming east, as compared with the larger fan and a very high prevailing southwest wind going west, probably accounts for some of this change in water temperature. Steps have already been taken to add 1" to the top of the present core, and this, along with an 18" fan, which can probably be quieted without any great sacrifice in efficiency, should cool this car satisfactorily.

CARBURETOR AND FUEL SYSTEM A pronounced and continuous pop-back in the muffler on deceleration was materially reduced by re-setting the idle adjustment. Incidentally, the idling adjustments are very inaccessible, and the screws should be slotted to permit adjusting with a screwdriver. Carburetion is still not entirely satisfactory, and just as soon as the car is available, we should have the carburetor representative work out the details. Vapor lock was a source of continuous trouble, particularly in starting after the engine was shut off following a hard run. To save time, ice packs and cold water were administered to the fuel pump, which relieved the situation immediately. On our arrival in L.A. Mr. Ames suggested installing a two-way electric impulse pump, and this was done, mounting the unit on the dash, where it was open to air flow. The first night out on the return trip was fairly cool, and no trouble was experienced. However, hot weather was experienced east of Salt Lake, and the impulse pump failed completely while driving at higher speeds, in addition to stalling when the engine was shut off. After considerable delay the mechanical pump was again connected, and no trouble was experienced with vapor lock in high speed driving. However, the same difficulty in starting after the engine was shut off was encountered. The electric impulse pump also materially affected radio reception. The present mechanical pump sits directly behind the distributor, restricting the air flow to a point where there is practically no circulation. This pump can be readily offset 45° to the right or left, which will place it in an air stream and unquestionably reduce the temperature. Lycoming are now investigating the possibility of making this change. This condition is a very serious one, and must be corrected.

MUFFLER AND EXHAUST SYSTEM Both exhaust manifolds were replaced twice, and the adjusting nuts taken up at regular intervals to eliminate exhaust leaks. It is questionable whether this construction will prove out satisfactorily without flaring the exhaust pipe to provide a seat for the packing. The exhaust pipe clamp bracket at the dash was not flexibly mounted, but this, I understand, has already been corrected. The exhaust pipe was riding the lower edge of the hole in body cross member, and providing more clearance eliminated considerable roughness and vibration in the engine. The exhaust pipe was wrapped with short pieces of asbestos, and this showed an improvement in quietness of the exhaust system. The muffler has an objectionable roar at 30-35 m.p.h., and is open for further improvement. The power roar of the engine is another point open for improvement, but it is questionable whether anything can be done, due to the limited amount of clearance and space for the carburetor silencer.

BATTERY Battery was completely run down after several days at L.A., and replaced with a 17-plate. No further battery trouble experienced on return trip. Success or failure of the car depends to a great extent upon positive and dependable operation of the electric gear shift, and Mr. Cord strongly recommends the use of a 21-plate battery to minimize any possibility of battery failure.

ENGINE Pronounced detonation at all speeds slightly improved by use of Ethyl gas. Carbon removed at L.A., and this reduced detonation temporarily. Inspection at L.A. showed main bearings, rod bearings, pistons and valves in very good condition. However, there was a trace of oil on top of pistons and excessive carbon accumulation considering mileage. Engine is equipped with three-ring Ray Day pistons and in view of the oil consumption record it is recommended that we change to a four-ring piston, which we know from past experience has proven very satisfactory. Engine operation smooth and with no outstanding periods. Valve mechanism also quiet. Engine sounds much quieter and detonation materially reduced at higher altitudes, with apparently no appreciable drop off in performance. Believe it advisable to conduct actual road tests with slightly lower compression ratio to determine effect on performance. No. 1, 2 and 3 plugs on right bank were replaced due to fouling.

BODY Upper and lower dovetails on all doors not contacting properly, with the result that after approximately 2,000 miles looseness developed in hinges and noise in door lock bolts and striker, due to excessive movement. Water leaks at windshield wipers and cowl ventilator. Noticeable side sway at rear end of body is observed through rear vision mirror. The front cushion could be lowered one inch, to provide more vision along top edge of shield, and without seriously affecting visibility over cowl. Sun visor and brackets seriously interfered with head clearance and were removed. Visor too deep vertically and does not shield sun and still permit vision straight ahead. Suggest reducing width to approximately one-half of present type. With the exception of noise caused by loose doors, body construction otherwise very quiet and extremely rigid and solid. Inside door and regulator handles very loose. This by all means must be corrected. Rear seat very comfortable. However, after long ride lack of sufficient leg room is felt. Metal clips which retain felt channels in top of doors rattle; additional clips would correct this condition. The maroon broadcloth fades very badly.

ELECTRIC GEAR SHIFT Failed to engage several times in first and third gear and investigation found cross shaft shift lever striking radiator pan and lever loose on shaft. Shift also failed to engage in fourth speed several times after continuous hard driving, which may have been caused by excessive heat. Present shift arrangement not entirely satisfactory, particularly for night driving, as it is somewhat difficult to readily select the desired speed. On several occasions the shift lever was also shifted back into neutral by accidentally hitting with the hand. Shifting from fourth to third speed is done quite frequently in passing another car, and to make a quick shift it would be more desirable to have fourth speed directly opposite third—that is, the same relation as the present second and third speeds. While in L.A. the shift failed to operate entirely on several occasions, and this may have been caused by low battery, as no further trouble was experienced with shift on return trip after 17-plate battery was installed. No heavy rains or muddy roads were encountered on trip, consequently no reliable information was obtained on effect on the shift under these conditions.

HEADLAMPS Source of continuous trouble on entire trip. Looseness in mounting brackets source of rattles. Cables broke, necessitating blocking the lamps for night driving. 32 C. P. bulbs very poor for high speed driving. Installation of 50 C. P. gave excellent lights. Entire layout very unsatisfactory. Engineering now working on several new applications.

MISCELLANEOUS Clutch pedal interferes with fender apron on extreme lower end of stroke. Startix cable broken in two. Very difficult to insert key in steering gear ignition lock. Ignition lock lever interferes with driver's knee and unknowingly is turned on and off. Rigid type radio antenna bracket entirely too low, causing serious interference over normal ramps. Flexible rubber mounting similar to present type will correct this condition. Paint on lower end of rear fenders badly pitted, due to throwing of gravel from front wheels. A chrome plated protecting shield similar to that used on speedster should be added. Rubber weather seal around trunk lid and doors torn loose. A metal retaining strip or some other positive means of anchoring the rubber should be used, instead of depending on rubber cement. In going down steep grades gasoline supply is exhausted, even with five to seven gallons in tank, due to unusually shallow tank. Mr. Thomas has already taken steps to correct this condition. Lubricant leaked out of transmission cover and inner end of joint, necessitating refilling. Lubricant works through front hub spline and onto brake shoes. Front tires strike headlamp housing and fender iron. Rear spring clip bolts strike body under frame. Ignition wire conduit brackets too close to cylinder head bolts. Lower left rear cylinder head bolt interferes with steering column, necessitating installing this bolt in the cylinder head before seating on the block. The visibility of the most important instrument, namely the speedometer, is obstructed by the steering wheel. This should be mounted in the present oil gauge position. Very important that cap on oil filler, located in cowl, is air-tight, as any leakage of oil fumes is readily drawn into the car with the windshield open. Sample Defiance electric air horns very unsatisfactory for tone. These were removed in L.A., and a set of our regular Delco electric installed. Radio very poor reception, probably due to over-tones and vibration in speaker. Steering wheel loose on column. All front fender iron bolts loose. Chrome wheel discs difficult to remove without damaging. Reset on clock inaccessibly located. Should be placed on outer face. Rear end of front seat adjuster channel damages shoes of rear seat passengers. Sharp surfaces should be rounded off or a rubber protecting cap added. Radiator hose sets too close to exhaust pipes. Forty pounds air pressure in front tires reduces tire squeal but results in harshness and shake. Radiator tie rod loose at dash. Bayonet oil gauge on crankcase very inaccessible. Should be extended higher up. Front seat adjuster mechanism rattles. Need for a toggle grip or grab handle of some kind keenly felt after riding long distance in rear seat. Arm rest would also add to comfort. Metal bottom in trunk compartment buckles and should be stiffened with additional reinforcing ribs. Oil filler cap difficult to pull up tight. Also, usual oil container used by filling stations cannot be readily used without spilling oil over cowl. Larger diameter filler spout would improve this condition. Rumble and buckle in gas tank. This corrected by installing a felt pad across center of top of tank. Clutch chatters.

GENERAL COMMENTS High gear performance under 30-35 m.p.h. sluggish but this can be expected with 2.75 ratio. Third speed performance good for driving in traffic and passing cars. Unable to comment on second gear, due to jumping out. Top speed in high gear showed 99 m.p.h. speedometer reading, which is probably actual car speed of approximately 92 m.p.h. Due to low, racy design public immediately associate car with high speeds, and top speed should at least equal or better our Supercharged eight performance. General comments on design, appearance and appointments highly favorable and public interest tremendous.

APPENDIX III

A Sampling of Cord Advertising

A sampling of major Cord ads are reproduced on the following pages. There were other ads that were variations on some of these, and some smaller ads that are not shown.

Many of the Cord ads shown here depict E306 #2. In some its non-production details are retouched. Other photographs used in the ads were taken at Cordhaven in November 1935, probably of cars that had been sent west for the Los Angeles auto show. The only production car shown in ad photographs is a supercharged sedan. (Unless otherwise noted, the original ads are from the author's collection.)

This is the first ad for the Cord 810. The car is E306 #2, and the shot was taken at Cordhaven. In this picture the inboard headlights, grooved bumper and Auburn bumper guards are still visible. The same ad was used several times before and after the November 1935 shows, sometimes to attract new dealers. Text was modified to suit the timing and purpose.

This ad features the Convertible Phaeton Sedan at Cordhaven. The man standing next to the showcar is Jack, the Cords' gardener.

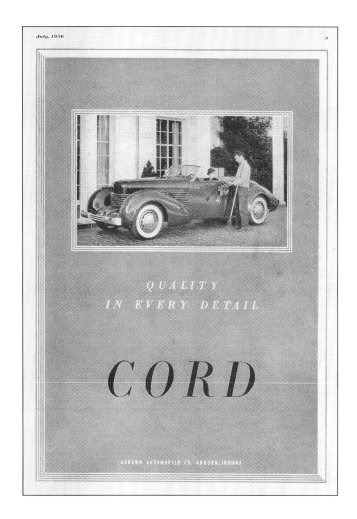

The first of a series featuring non-photographic art. Auburn suggested that wealthy parents buy their daughters a Cord.

241

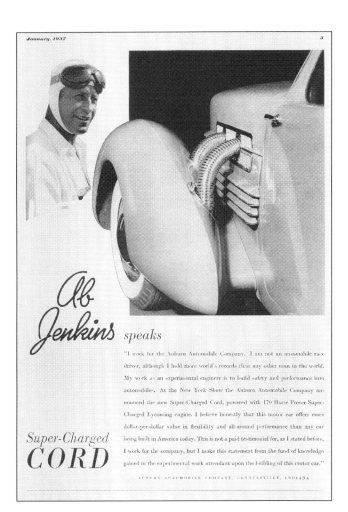

One of the few ads in which the horsepower of the Super-Charged Cord is mentioned. As we've seen, the figure was not correct.

One of many ads placed by R.S.M. Automobiles Ltd., Auburn's exclusive dealer in London. The car is a showcar photographed at Cordhaven. RSM's ad agency moved the steering wheel to the right. (Ron Irwin)

The Cord's special image encouraged vendors to associate themselves with it in their own advertising. This ad for piston rings was one of a series.

Possibly the most famous Cord ad, because of its oft-repeated slogan.

APPENDIX iv

Numbers

Three numbers are used to identify the Model 810 and 812 Cords: serial, engine and body.

A number plate screwed to the right side of the cowl lists both serial and engine numbers. (On right hand drive cars for export this plate was on the left side of the cowl.) Below this, on production cars, is another smaller plate giving the body number. (On the showcars, this plate may be located under the front seat.)

Here's how to read the numbers. A single fictitious serial number will be used as illustration.

Model 810

Serial number. All 810s are 1936 models. The serial number contains four digits, plus a letter. The number was based on the frame number, which was stamped into the right side of each stub frame. 1,000 was added to the frame number to arrive at the serial number. 1,001 to 1,100 inclusive were assigned to the showcars. Production models began with 1,101.

The letter after the four digits indicates the body type, as follows:

A Five-passenger Westchester sedan

F Two-passenger convertible coupe

H Five-passenger Convertible Phaeton Sedan

M Coupe

S Four-passenger Beverly sedan

(Example: a 1936 Beverly sedan with frame number 440 would have serial number 810 1440S.)

Engine number. The engine number is stamped on the engine block, on a boss at the left front. The same number is stamped on the cowl plate. All unsupercharged engines use the prefix "FB."

Model 812

All 1937 models are 812, but not all 812 cars were originally built as 1937 cars. As many as 125 unsold production 810s at the factory and 300 in the hands of factory branches were renumbered as 812s in September 1936. The renumbered 810s were given new 812 serial numbers between 1,001 and 1,525. Renumbered plates for cars that were not at the factory were sent to each district sales manager. He saw that the plates were changed, and returned the old plates to the factory where records of the changes were kept.

So all 812s with serial numbers of 1,525 or lower were originally 810s. The first car built as an 812 was numbered 1,526. On these cars there is no correlation between the frame numbers and the serial numbers.

Some 810 showcars were renumbered as 812s too. On these showcars the original four-digit serial numbers were retained; just the prefix was changed from 810 to 812.

Serial number. The serial number is the frame number plus 1,000. The frame number on the long-wheelbase Custom series has the digit "9" added as the first digit.

Examples:
A Custom Beverly with frame number 9,440 would be 812 10440S.
A Convertible Coupe with frame number 440 would be 812 1440F

The model designation 812 has nothing to do with whether or not the car was supercharged. Supercharged cars were designated by the digit "3" added in front of the serial number.

Examples:
A supercharged Beverly with frame number 440 would have serial number 812 31440S.
A supercharged Custom Beverly would be 812 310440S.

As with the 810, a letter after the serial number indicates the body type:

A Five-passenger Westchester sedan

B Berline

F Two-passenger convertible coupe

H Five-passenger Convertible Phaeton Sedan

K We don't know the reason for this letter. It can be found on two right hand drive Westchester sedans in Australia.

M Coupe

S Five-passenger Beverly sedan, or Custom Beverly

> **About Frame Numbers**
>
> Frame numbers started with "1." The number was stamped in digits about 1/2" high. The impression was so light that a single coat of paint sometimes obscured it.
>
> Later open cars had reinforced stub frames intended to reduce cowl shake. These frames had no number at all.

Engine number. Unsupercharged engines were prefixed "FB." This is true whether the fuel pump was mounted on top of the intake manifold, or at the left rear of the block as on the later engines. Supercharged engines were prefixed "FC."

810 and 812 Body numbers. This small plate includes the words "CENTRAL MANUFACTURING CO." The body number includes the prefix "C," followed by a body type number. The next line on the plate listed the body number.

The term "fastback" and "bustle trunk" as used below are intended to be descriptive, and are not nomenclature used by Auburn. Here's what the body type numbers mean:

C90 Sedan with fastback

C91 Convertible Phaeton Sedan

C92 Convertible Coupe

C94 Beverly sedan, with fixed armrest seats

C95 Berline with fastback, possibly a prototype

C96 Sedan with bustle trunk

C98 Convertible coupe prototype

C99 Two-passenger coupe

C100 Phaeton with bustle trunk

C101 Custom Beverly six-window sedan, 132" wheelbase

C102 Convertible Coupe with rumble seat

C103 Custom Berline, 132" wheelbase

C105 Custom Beverly, 132" wheelbase

C105 Custom Berline, 135" wheelbase

The list below reconstructs the known original number combinations as the cars left the factory. They represent decades of research by Ronald B. Irwin, 810-812 Cord Historian of the Auburn Cord Duesenberg Club. The numbers are based on information submitted by owners. While every attempt has been made to insure accuracy, no guarantees are made or should be assumed.

Known facts about individual cars are given next to their numbers.

1936 Model 810

Serial #		Engine #		Body #			Model	Body Type	Interior pattern	Comments
				C	90	109	Westchester	Sedan (fastback)	Plain	Showcar, handbuilt for the 1935 shows.
810 1014	A	FB	1911				Westchester	Sedan (fastback)	Plain	Showcar, handbuilt for the 1935 shows.
810 1017	A	FB	1825				Westchester	Sedan (fastback)	Plain	Showcar, handbuilt for the 1935 shows.
810 1021	A			C	90	128	Westchester	Sedan (fastback)	Plain	Showcar, handbuilt for the 1935 shows. Had the Coppertone color scheme.
810 1024	A	FB	1709	C	90	116	Westchester	Sedan (fastback)	Plain	Showcar, handbuilt, probably in time for the 1935 shows
810				C	91	10?	Convertible Phaeton			The only surviving convertible show car.
810 1035	A	FB	1291				Westchester	Sedan (fastback)	Plain	Showcar, handbuilt for the 1935 shows or soon afterward.
810 1038	A	FB	1846	C	90	131	Westchester	Sedan (fastback)	Plain	Showcar, handbuilt for the 1935 shows or soon afterward.
810 1042	A	FB	258				Westchester	Sedan (fastback)	Plain	Showcar, handbuilt, probably after the 1935 shows.
810 1044	A	FB	1822	C	90	136	Westchester	Sedan (fastback)	Plain	Showcar, handbuilt, probably after the 1935 shows.
810				C	90	147	Westchester	Sedan (fastback)	Plain	Showcar, handbuilt, probably after the 1935 shows.
810				C	90	160	Westchester	Sedan (fastback)	Plain	Showcar, handbuilt, probably after the 1935 shows.
810 1094	A	FB	240	C	90	167	Westchester	Sedan (fastback)	Plain	Showcar, handbuilt, probably after the 1935 shows.
810 1101	A	FB	241				Westchester	Sedan (fastback)	Plain	The first production Cord 810, dismantled by Auburn for evaluation.
810 1103	A						Westchester	Sedan (fastback)	Plain	
810 1105	A						Westchester	Sedan (fastback)	Plain	General Motors bought this car to evaluate.
810 1107	A	FB	246				Westchester	Sedan (fastback)	Plain	
810 1108	A	FB	266				Westchester	Sedan (fastback)	Plain	
810 1109	A	FB	250	C	90	184	Westchester	Sedan (fastback)	Plain	
810 1110	S	FB	245				Beverly	Sedan (fastback)	Fixed armrests, front and rear	
810 1111	A	FB	880	C	90	185	Westchester	Sedan (fastback)	Plain	
810 1113	A	FB	244	C	90	187	Westchester	Sedan (fastback)	Plain	
810 1114	A	FB	228	C	90	188	Westchester	Sedan (fastback)	Plain	
810 1118	A	FB	249				Westchester	Sedan (fastback)	Plain	
810 1120	A	FB	905	C	90	191	Westchester	Sedan (fastback)	Plain	
810 1121	S	FB	747				Beverly	Sedan (fastback)	Fixed armrests, front and rear	
810 1122	A	FB	313	C	90	200	Westchester	Sedan (fastback)	Plain	
810 1123	A	FB	323	C	90	196	Westchester	Sedan (fastback)	Plain	
810 1125	S	FB	322	C	94	105	Beverly	Sedan (fastback)	Fixed armrests, front and rear	Early Beverlys apparently carried a C 94 body number.
810 1130	A	FB	297	C	90	208	Westchester	Sedan (fastback)	Plain	
810 1140	A	FB	243	C	90	220	Westchester	Sedan (fastback)	Plain	
810 1149	A	FB	314	C	90	224	Westchester	Sedan (fastback)	Plain	
810		FB	316				Westchester	Sedan (fastback)	Plain	
810 1151	A	FB	318	C	90	225	Westchester	Sedan (fastback)	Plain	
810 1154	S	FB	269	C	90	210	Beverly	Sedan (fastback)	Fixed armrests, front and rear	
810 1157	A	FB	457	C	90	233	Westchester	Sedan (fastback)	Plain	
810 1159	A	FB	252				Westchester	Sedan (fastback)	Plain	
810 1164	A	FB	451	C	90	247	Westchester	Sedan (fastback)	Plain	
810 1169	A	FB	276	C	90	203	Westchester	Sedan (fastback)	Pleated leather	Right hand drive.
810 1176	A	FB	282	C	90	240	Westchester	Sedan (fastback)	Plain	
810 1179	A						Westchester	Sedan (fastback)	Plain	
810 1184	S	FB	372	C	90	242	Beverly	Sedan (fastback)	Folding front armrest, fixed rear	
810 1191	A						Westchester	Sedan (fastback)	Plain	
810 1192	A	FB	369	C	90	287	Westchester	Sedan (fastback)	Plain	
810 1193	A	FB	434	C	90	253	Westchester	Sedan (fastback)	Plain	
810 1197	A	FB	351				Westchester	Sedan (fastback)	Plain	
810 1200	A						Westchester	Sedan (fastback)	Plain	
810 1205	A	FB	340				Westchester	Sedan (fastback)	Plain	Exported to Canada.
810 1209	A	FB	225	C	90	295	Westchester	Sedan (fastback)	Plain	
810 1211	A	FB	339	C	90	294	Westchester	Sedan (fastback)	Plain	
810 1217	A						Westchester	Sedan (fastback)	Plain	
810 1222	A	FB	486	C	90	305	Westchester	Sedan (fastback)	Plain	
810 1223	S	FB	405	C	90	291	Beverly	Sedan (fastback)	Fixed armrests, front and rear	
810 1223	A	FB	436				Westchester	Sedan (fastback)	Plain	
810 1228	A	FB	1225	C	90	303	Westchester	Sedan (fastback)	Plain	May be A-C-D Company-installed engine.
810 1233	A	FB	565				Westchester	Sedan (fastback)	Plain	
810 1237	A	FB	408	C	90	316	Westchester	Sedan (fastback)	Plain	
810 1238	A	FB	536				Westchester	Sedan (fastback)	Plain	
810 1240	A	FB	328				Westchester	Sedan (fastback)	Plain	
810		FB	307	C	90	344	Westchester	Sedan (fastback)	Plain	Right hand drive.
810 1243	A	FB	396				Westchester	Sedan (fastback)	Plain	
810 1258	S	FB	395	C	90	320	Beverly	Sedan (fastback)	Fixed armrests, front and rear	
810 1263	A	FB	543	C	90	349	Westchester	Sedan (fastback)	Plain	
810 1269	A	FB	501				Westchester	Sedan (fastback)	Plain	
810 1279	A	FB	393				Westchester	Sedan (fastback)	Plain	
810 1280	S	FB	274				Beverly	Sedan (fastback)	Fixed armrests, front and rear	
810 1282	A	FB	735	C	90	372	Westchester	Sedan (fastback)	Plain	
810 1283	A	FB	722	C	90	384	Westchester	Sedan (fastback)	Plain	
810 1284	A	FB	1576	C	90	179	Westchester	Sedan (fastback)	Plain	May be A-C-D Company-installed engine.
810 1289	A						Westchester	Sedan (fastback)	Plain	
810 1291	S	FB	357	C	90	340	Beverly	Sedan (fastback)	Fixed armrests, front and rear	
810 1302	A	FB	450				Westchester	Sedan (fastback)	Plain	
810 1312	A						Westchester	Sedan (fastback)	Plain	

Serial #		Engine #		Body #			Model	Body Type	Interior pattern	Comments
810	1316	A	FB	414			Westchester	Sedan (fastback)	Plain	
810	1321	A	FB	421			Westchester	Sedan (fastback)	Plain	
810	1325	A					Westchester	Sedan (fastback)	Plain	
810	1327	A	FB	507	C 90	414	Westchester	Sedan (fastback)	Plain	
810	1333	A	FB	562	C 90	432	Westchester	Sedan (fastback)	Pleated leather	
810	1334	A					Westchester	Sedan (fastback)	Plain	
810	1336	S	FB	564	C 90	312	Beverly	Sedan (fastback)	Fixed armrests, front and rear	
810	1337	A	FB	520	C 90	413	Westchester	Sedan (fastback)	Plain	
810	1339	A	FB	567	C 90	434	Westchester	Sedan (fastback)	Plain	
810	1341	A	FB	570			Westchester	Sedan (fastback)	Plain	Exported to Canada.
810	1342	A	FB	571	C 90	441	Westchester	Sedan (fastback)	Plain	
810	1344	A	FB	569			Westchester	Sedan (fastback)	Plain	
810	1345	A	FB	469	C 90	428	Westchester	Sedan (fastback)	Plain	
810					C 90	438	Beverly	Sedan (fastback)	Fixed armrests, front and rear	
810	1347	A	FB	579	C 90	458	Westchester	Sedan (fastback)	Plain	
810	1348	A	FB	577	C 90	452	Westchester	Sedan (fastback)	Plain	
810	1351	A	FB	635			Westchester	Sedan (fastback)	Plain	
810	1354	A	FB	2121			Westchester	Sedan (fastback)	Plain	May be A-C-D Company-installed engine.
810	1356	A	FB	479			Westchester	Sedan (fastback)	Plain	
810	1358	A	FB	438	C 90	464	Westchester	Sedan (fastback)	Plain	Exported to Canada.
810	1361	A	FB	576			Westchester	Sedan (fastback)	Plain	
810	1368	A	FB	575	C 90	470	Westchester	Sedan (fastback)	Plain	
810	1369	A	FB	631	C 90	453	Westchester	Sedan (fastback)	Plain	
810	1370	A	FB	630	C 90	481	Westchester	Sedan (fastback)	Plain	
810			FB	671			Westchester	Sedan (fastback)	Plain	
810	1372	A	FB	513	C 90	476	Westchester	Sedan (fastback)	Plain	
810	1373	A	FB	674	C 90	475	Westchester	Sedan (fastback)	Plain	
810	1381	A	FB	471	C 90	483	Westchester	Sedan (fastback)	Plain	
810	1383	A	FB	480	C 90	497	Westchester	Sedan (fastback)	Plain	
810	1387	S	FB	586	C 90	451	Beverly	Sedan (fastback)	Fixed armrests, front and rear	Right hand drive.
810	1388	S					Beverly	Sedan (fastback)	Fixed armrests, front and rear	Right hand drive.
810	1389	A	FB	484	C 90	485	Westchester	Sedan (fastback)	Pleated leather	Bolt-on accesory trunk.
810	1393	A	FB	681			Westchester	Sedan (fastback)	Plain	
810	1398	A	FB	492	C 90	496	Westchester	Sedan (fastback)	Plain	
810	1402	A	FB	664	C 90	443	Westchester	Sedan (fastback)	Plain	
810	1403	S	FB	663	C 90	399	Beverly	Sedan (fastback)	Fixed armrests, front and rear	
810	1405	A	FB	452	C 90	469	Westchester	Sedan (fastback)	Plain	
810	1408	A	FB	487			Westchester	Sedan (fastback)	Plain	
810	1415	A	FB	462	C 90	508	Westchester	Sedan (fastback)	Plain	
810	1418	A	FD	437	C 90	513	Westchester	Sedan (fastback)	Plain	
810	1422	A	FB	657	C 90	507	Westchester	Sedan (fastback)	Plain	
810	1425	A	FB	667	C 90	519	Westchester	Sedan (fastback)	Plain	
810	1427	S	FB	668			Beverly	Sedan (fastback)	Fixed armrests, front and rear	
810	1433	A	FB	625			Westchester	Sedan (fastback)	Plain	Exported to Canada.
810	1435	S	FB	502	C 90	427	Beverly	Sedan (fastback)	Fixed armrests, front and rear	
810	1437	S	FB	615	C 90	707	Beverly	Sedan (fastback)	Fixed armrests, front and rear	
810	1439	A	FB	636			Westchester	Sedan (fastback)	Plain	
810					C 90	524	Westchester	Sedan (fastback)	Plain	
810	1445	A	FB	649	C 90	533	Westchester	Sedan (fastback)	Plain	
810	1451	A	FB	655			Westchester	Sedan (fastback)	Plain	
810	1452	S	FB	660			Beverly	Sedan (fastback)	Fixed armrests, front and rear	
810	1457	S					Beverly	Sedan (fastback)	Fixed armrests, front and rear	
810	1460	S	FB	605	C 90	529	Beverly	Sedan (fastback)	Fixed armrests, front and rear	
810	1463	A	FB	703			Westchester	Sedan (fastback)	Plain	
810	1464	A	FB	706	C 90	550	Westchester	Sedan (fastback)	Plain	
810	1468	A	FB	254	C 90	555	Westchester	Sedan (fastback)	Plain	
810	1469	A	FB	713			Westchester	Sedan (fastback)	Plain	
810	1476	A	FB	602	C 90	563	Westchester	Sedan (fastback)	Plain	
810	1477	A	FB	603	C 90	559	Westchester	Sedan (fastback)	Plain	
810			FB	608			Westchester	Sedan (fastback)	Plain	
810	1480	A	FB	532			Westchester	Sedan (fastback)	Plain	
810	1482	A	FB	516	C 90	585	Westchester	Sedan (fastback)	Plain	
810	1485	A	FB	736	C 90	586	Westchester	Sedan (fastback)	Plain	
810	1486	A	FB	729			Westchester	Sedan (fastback)	Plain	This car was redesigned into The Phantom Corsair.
810	1490	A	FB	732	C 90	580	Westchester	Sedan (fastback)	Plain	
810	1491	A					Westchester	Sedan (fastback)	Plain	
810	1496	A	FB	730	C 90	574	Westchester	Sedan (fastback)	Plain	
810	1497	S	FB	721	C 90	404	Beverly	Sedan (fastback)	Fixed armrests, front and rear	
810	1498	A	FB	725			Westchester	Sedan (fastback)	Plain	
810	1506	S	FB	619	C 90	487	Beverly	Sedan (fastback)	Fixed armrests, front and rear	
810			FB	679			Beverly	Sedan (fastback)	Fixed armrests, front and rear	
810	1511	A	FB	692	C 90	657	Westchester	Sedan (fastback)	Plain	
810	1512	A	FB	647	C 90	614	Westchester	Sedan (fastback)	Plain	
810	1513	A	FB	707	C 90	680	Westchester	Sedan (fastback)	Plain	
810	1514	A	FB	678	C 90	628	Westchester	Sedan (fastback)	Plain	Right hand drive.
810	1516	A	FB	634	C 90	666	Westchester	Sedan (fastback)	Plain	
810	1517	A	FB	704	C 90	673	Westchester	Sedan (fastback)	Plain	
810	1518	A					Westchester	Sedan (fastback)	Plain	
810	1522	A	FB	780	C 90	661	Westchester	Sedan (fastback)	Plain	

Serial #		Engine #		Body #			Model	Body Type	Interior pattern	Comments
810	1525	A	FB	788	C	90	667	Westchester	Sedan (fastback)	Plain	
810	1527	S	FB	787	C	90	481	Beverly	Sedan (fastback)	Fixed armrests, front and rear	
810	1531	A						Westchester	Sedan (fastback)	Plain	
810	1536	A	FB	758				Westchester	Sedan (fastback)	Plain	
810	1539	A	FB	767	C	90	615	Westchester	Sedan (fastback)	Plain	
810	1540	S	FB	792	C	90	710	Beverly	Sedan (fastback)	Fixed armrests, front and rear	
810	1541	S	FB	699				Beverly	Sedan (fastback)	Fixed armrests, front and rear	Probably right hand drive.
810	1542	S	FB	805	C	90	712	Beverly	Sedan (fastback)	Fixed armrests, front and rear	
810	1544	A	FB	813	C	90	709	Westchester	Sedan (fastback)	Plain	
810	1546	A	FB	795	C	90	676	Westchester	Sedan (fastback)	Plain	
810	1548	A	FB	688				Westchester	Sedan (fastback)	Plain	Right hand drive.
810	1552	A	FB	810				Westchester	Sedan (fastback)	Plain	
810	1554	A	FB	808	C	90	730	Westchester	Sedan (fastback)	Plain	
810	1556	A	FB	503				Westchester	Sedan (fastback)	Plain	
810	1561	S	FB	740	C	90	704	Beverly	Sedan (fastback)	Fixed armrests, front and rear	
810	1562	A	FB	724	C	90	731	Westchester	Sedan (fastback)	Plain	
810	1564	A	FB	478				Westchester	Sedan (fastback)	Plain	
810	1571	A	FB	759	C	90	764	Westchester	Sedan (fastback)	Plain	
810	1574	S	FB	769	C	90	775	Beverly	Sedan (fastback)	Fixed armrests, front and rear	
810	1575	A	FB	762	C	90	754	Westchester	Sedan (fastback)	Plain	
810			FB	771				Westchester	Sedan (fastback)	Plain	
810	1579	A	FB	774	C	90	590	Westchester	Sedan (fastback)	Plain	
810	1580	A	FB	371				Westchester	Sedan (fastback)	Plain	
810	1581	A	FB	498	C	90	595	Westchester	Sedan (fastback)	Plain	
810	1582	A	FB	514	C	90	596	Westchester	Sedan (fastback)	Plain	
810	1584	A	FB	493				Westchester	Sedan (fastback)	Plain	
810	1585	A						Westchester	Sedan (fastback)	Plain	
810	1586	A	FB	582	C	90	592	Westchester	Sedan (fastback)	Plain	Exported to Canada.
810	1587	A	FB	820				Westchester	Sedan (fastback)	Plain	Exported to Canada.
810	1589	S	FB	745				Beverly	Sedan (fastback)	Fixed armrests, front and rear	
810	1591	A	FB	287	C	90	694	Westchester	Sedan (fastback)	Plain	
810	1592	A	FB	1944				Westchester	Sedan (fastback)	Plain	May be A-C-D Company-installed engine.
810	1598	A	FB	464				Westchester	Sedan (fastback)	Plain	Exported to Canada.
810	1602	A	FB	1133	C	90	760	Westchester	Sedan (fastback)	Plain	Exported to Canada. May be A-C-D Company-installed engine.
810			FB	998				Westchester	Sedan (fastback)	Plain	
810	1609	A	FB	999	C	90	634	Westchester	Sedan (fastback)	Plain	
810					C	90	698	Westchester	Sedan (fastback)	Plain	
810	1622	A	FB	852	C	90	842	Westchester	Sedan (fastback)	Plain	
810	1630	A	FB	841	C	90	822	Westchester	Sedan (fastback)	Plain	
810	1631	A	FB	833				Westchester	Sedan (fastback)	Plain	
810	1633	S	FB	850				Beverly	Sedan (fastback)	Fixed armrests, front and rear	
810	1634	A	FB	2179				Westchester	Sedan (fastback)	Plain	May be A-C-D Company-installed engine.
810	1644	A	FB	837	C	90	796	Westchester	Sedan (fastback)	Plain	
810	1645	A	FB	825	C	90	814	Westchester	Sedan (fastback)	Plain	
810	1646	A	FB	826				Westchester	Sedan (fastback)	Plain	
810	1649	A	FB	568	C	90	606	Westchester	Sedan (fastback)	Plain	
810	1654	A						Westchester	Sedan (fastback)	Plain	
810	1655	A	FB	915				Westchester	Sedan (fastback)	Plain	
810	1659	S	FB	903	C	90	871	Beverly	Sedan (fastback)	Folding front armrest, fixed rear	
810	1660	A	FB	912				Westchester	Sedan (fastback)	Plain	
810	1668	A	FB	898	C	90	882	Westchester	Sedan (fastback)	Plain	
810	1670	A	FB	882				Westchester	Sedan (fastback)	Plain	
810	1679	A	FB	904	C	90	774	Westchester	Sedan (fastback)	Plain	
810	1680	A	FB	902	C	90	763	Westchester	Sedan (fastback)	Plain	
810	1684	A	FB	869	C	90	827	Westchester	Sedan (fastback)	Plain	
810	1689	A	FB	880	C	90	806	Westchester	Sedan (fastback)	Plain	
810	1690	A	FB	858				Westchester	Sedan (fastback)	Plain	
810	1693	A	FB	879	C	90	797	Westchester	Sedan (fastback)	Plain	
810	1695	A	FB	867	C	90	773	Westchester	Sedan (fastback)	Plain	
810	1696	A	FB	863	C	90	839	Westchester	Sedan (fastback)	Plain	
810			FB	868				Westchester	Sedan (fastback)	Plain	
810	1700	A						Westchester	Sedan (fastback)	Plain	
810	1701	A	FB	875	C	90	884	Westchester	Sedan (fastback)	Plain	
810	1702	A	FB	859				Westchester	Sedan (fastback)	Plain	
810	1704	A	FB	870	C	90	600	Westchester	Sedan (fastback)	Plain	
810	1705	S	FB	872	C	90	605	Beverly	Sedan (fastback)	Fixed armrests, front and rear	
810	1706	A	FB	862	C	90	611	Westchester	Sedan (fastback)	Plain	
810	1710	A	FB	885	C	90	784	Westchester	Sedan (fastback)	Plain	
810	1712	A	FB	856				Westchester	Sedan (fastback)	Plain	
810	1716	A	FB	871	C	90	916	Westchester	Sedan (fastback)	Plain	
810	1718	A	FB	411				Westchester	Sedan (fastback)	Plain	
810	1721	A	FB	943	C	90	915	Westchester	Sedan (fastback)	Plain	
810	1724	A	FB	963	C	90	762	Westchester	Sedan (fastback)	Plain	
810	1727	S	FB	957	C	90	782	Beverly	Sedan (fastback)	Fixed armrests, front and rear	
810	1728	A	FB	906	C	90	877	Westchester	Sedan (fastback)	Plain	
810	1732	A	FB	960	C	90	800	Westchester	Sedan (fastback)	Plain	
810	1734	A	FB	1022				Westchester	Sedan (fastback)	Plain	
810	1736	A	FB	1027				Westchester	Sedan (fastback)	Plain	

Serial #	Engine #	Body #	Model	Body Type	Interior pattern	Comments
810		C 90 815	Beverly	Sedan (fastback)	Fixed armrests, front and rear	
810 1749 A	FB 990		Westchester	Sedan (fastback)	Plain	
810 1761 A	FB 900		Westchester	Sedan (fastback)	Plain	
810 1766 A	FB 896	C 90 701	Westchester	Sedan (fastback)	Plain	
810 1769 A	FB 951	C 90 705	Westchester	Sedan (fastback)	Plain	
810 1770 A	FB 865	C 90 728	Westchester	Sedan (fastback)	Plain	
810 1771 A	FB 952	C 90 732	Westchester	Sedan (fastback)	Plain	
810 1772 A	FB 1633	C 90 731	Westchester	Sedan (fastback)	Plain	May be A-C-D Company-installed engine.
810 1776 A	FB 972		Westchester	Sedan (fastback)	Plain	
810 1779 A	FB 985		Westchester	Sedan (fastback)	Plain	
810 1783 A	FB 956	C 90 767	Westchester	Sedan (fastback)	Plain	
810	FB 971		Westchester	Sedan (fastback)	Plain	
810 1787 A	FB 975	C 90 755	Westchester	Sedan (fastback)	Plain	
810 1789 A	FB 974	C 90 721	Westchester	Sedan (fastback)	Plain	
810 1792 A	FB 981	C 90 717	Westchester	Sedan (fastback)	Plain	
810 1793 A	FB 978	C 90 725	Westchester	Sedan (fastback)	Plain	
810 1795 A	FB 979	C 90 601	Westchester	Sedan (fastback)	Plain	
810 1798 A	FB 1013	C 90 723	Westchester	Sedan (fastback)	Plain	
810 1802 A	FB 994	C 90 714	Westchester	Sedan (fastback)	Plain	
810 1807 A	FB 1005	C 90 651	Westchester	Sedan (fastback)	Plain	
810		C 90 787	Westchester	Sedan (fastback)	Plain	
810 1811 A	FB 1135		Westchester	Sedan (fastback)	Plain	
810 1814 A	FB 791		Westchester	Sedan (fastback)	Plain	
810 1815 A	FB 1004	C 90 639	Westchester	Sedan (fastback)	Plain	
810 1816 A	FB 1008	C 90 616	Westchester	Sedan (fastback)	Plain	
810 1818 S	FB 1112		Beverly	Sedan (fastback)	Fixed armrests, front and rear	
810 1819 A	FB 1020	C 90 632	Westchester	Sedan (fastback)	Plain	
810 1820 S	FB 996		Beverly	Sedan (fastback)	Fixed armrests, front and rear	
810 1821 A	FB 1025	C 90 655	Westchester	Sedan (fastback)	Plain	Exported to Canada.
810	FB 1034		Westchester	Sedan (fastback)	Plain	
810 1824 A	FB 1088		Westchester	Sedan (fastback)	Plain	
810 1827 A	FB 1086	C 90 626	Westchester	Sedan (fastback)	Plain	
810 1829 A	FB 1102	C 90 613	Westchester	Sedan (fastback)	Plain	
810 1830 A	FB 1016	C 90 624	Westchester	Sedan (fastback)	Plain	
810 1831 A	FB 1091	C 90 300	Westchester	Sedan (fastback)	Plain	
810 1832 A	FB 1090	C 90 996	Westchester	Sedan (fastback)	Plain	
810 1834 A	FB 1097	C 90 974	Westchester	Sedan (fastback)	Plain	
810 1836 A			Westchester	Sedan (fastback)	Plain	
810 1837 A	FB 1098	C 90 929	Westchester	Sedan (fastback)	Plain	
810 1839 A	FB 1101	C 90 1003	Westchester	Sedan (fastback)	Plain	
810 1847 A	FB 1084		Westchester	Sedan (fastback)	Plain	
810 1848 A	FB 1112		Westchester	Sedan (fastback)	Plain	
810 1850 A	FB 2211	C 90 1012	Westchester	Sedan (fastback)	Plain	May be A-C-D Company-installed engine.
810 1855 A	FB 496		Westchester	Sedan (fastback)	Plain	
810 1857 A	FB 422		Westchester	Sedan (fastback)	Plain	
810 1860 A			Westchester	Sedan (fastback)	Plain	
810 1869 A			Westchester	Sedan (fastback)	Plain	
810		C 90 1026	Westchester	Sedan (fastback)	Plain	
810 1875 A	FB 1137	C 90 1015	Westchester	Sedan (fastback)	Plain	
810 1876 A	FB 1109	C 90 1035	Westchester	Sedan (fastback)	Plain	Exported to Canada.
810 1879 A	FB 1129		Westchester	Sedan (fastback)	Plain	
810 1881 A	FB 1104	C 90 1002	Westchester	Sedan (fastback)	Plain	
810	FB 1136		Westchester	Sedan (fastback)	Plain	
810 1900 A	FB 1176		Westchester	Sedan (fastback)	Plain	
810 1903 A	FB 1175	C 90 974	Westchester	Sedan (fastback)	Plain	
810 1904 A	FB 1139	C 90 1000	Westchester	Sedan (fastback)	Plain	
810 1910 A	FB 1157	C 90 848	Westchester	Sedan (fastback)	Plain	
810 1912 A	FB 1215		Westchester	Sedan (fastback)	Plain	
810 1913 A	FB 1218	C 90 1065	Westchester	Sedan (fastback)	Plain	
810 1914 A	FB 1220	C 90 1049	Westchester	Sedan (fastback)	Plain	
810 1916 A	FB 1183	C 90 1051	Westchester	Sedan (fastback)	Plain	
810 1917 A	FB 1199		Westchester	Sedan (fastback)	Plain	
810		C 90 1050	Westchester	Sedan (fastback)	Plain	
810 1920 A	FB 1163	C 90 1067	Westchester	Sedan (fastback)	Plain	
810 1925 A	FB 1181	C 90 1062	Westchester	Sedan (fastback)	Plain	
810 1926 H	FB 1086		Convertible Phaeton			
810	FB 2417	C 91 120	Convertible Phaeton			May be A-C-D Company-installed engine.
810 1928 A	FB 1194		Westchester	Sedan (fastback)	Plain	
810 1934 A	FB 3292	C 90 966	Westchester	Sedan (fastback)	Plain	May be A-C-D Company-installed engine.
810 1939 A	FB 1190	C 90 925	Westchester	Sedan (fastback)	Plain	
810 1945 S	FB 1257	C 90 808	Beverly	Sedan (fastback)	Fixed armrests, front and rear	
810 1946 A	FB 1259		Westchester	Sedan (fastback)	Plain	
810 1947 A	FB 1217	C 90 1059	Westchester	Sedan (fastback)	Plain	
810 1949 A	FB 1246	C 90 1086	Westchester	Sedan (fastback)	Plain	
810 1953 A	FB 1238		Westchester	Sedan (fastback)	Plain	
810 1954 A	FB 1245	C 90 1083	Westchester	Sedan (fastback)	Plain	Exported to Canada.
810		C 90 1102	Westchester	Sedan (fastback)	Plain	
810 1962 H	FB 1221	C 91 122	Convertible Phaeton			
810 1963 A	FB 1214	C 90 1113	Westchester	Sedan (fastback)	Plain	

Serial #		Engine #		Body #		Model	Body Type	Interior pattern	Comments
810 1966	A			C 90	1063	Westchester	Sedan (fastback)	Plain	
810 1970	A	FB	1212			Westchester	Sedan (fastback)	Plain	
810 1971	A	FB	1195	C 90	1037	Westchester	Sedan (fastback)	Plain	Exported to Canada.
810 1975	H	FB	2582	C 91	121	Convertible Phaeton			May be A-C-D Company-installed engine.
810 1978	A	FB	1292			Westchester	Sedan (fastback)	Plain	
810				C 91	128	Convertible Phaeton			
810 1984	H	FB	385			Convertible Phaeton			
810 1985	H	FB	1274	C 91	133	Convertible Phaeton			
810 1986	H			C 91	127	Convertible Phaeton			
810 1988	A	FB	1267	C 90	963	Westchester	Sedan (fastback)	Plain	Exported to Canada.
810 1993	A	FB	1277	C 90	1109	Westchester	Sedan (fastback)	Plain	
810 1994	H	FB	1273	C 91	132	Convertible Phaeton			Exported to Canada.
810 1996	A	FB	1272	C 90	1122	Westchester	Sedan (fastback)	Plain	
810 2003	A	FB	232	C 90	1121	Westchester	Sedan (fastback)	Plain	
810 2005	H	FB	1263	C 91	126	Convertible Phaeton			
810 2006	A	FB	1260	C 90	1142	Westchester	Sedan (fastback)	Plain	
810 2015	A	FB	1289	C 90	1159	Westchester	Sedan (fastback)	Plain	
810 2016	A			C 90	1084	Westchester	Sedan (fastback)	Plain	
810 2018	A	FB	1250	C 90	1445	Westchester	Sedan (fastback)	Plain	
810 2019	A	FB	1302	C 90	1075	Westchester	Sedan (fastback)	Plain	
810 2023	A	FB	1301	C 90	1161	Westchester	Sedan (fastback)	Plain	
810 2025	A	FB	1294	C 90	1163	Westchester	Sedan (fastback)	Plain	
810 2027	A					Westchester	Sedan (fastback)	Plain	
810 2028	A	FB	3315	C 90	1156	Westchester	Sedan (fastback)	Plain	May be A-C-D Company-installed engine.
810 2031	A	FB	1167	C 90	1166	Westchester	Sedan (fastback)	Plain	
810 2032	A	FB	1165			Westchester	Sedan (fastback)	Plain	
810 2037	H	FB	1308	C 91	125	Convertible Phaeton			
810 2038	A	FB	1233	C 90	1169	Westchester	Sedan (fastback)	Plain	
810				C 90	1176				
810 2045	A	FB	1240	C 90	1139	Westchester	Sedan (fastback)	Plain	
810 2049	A	FB	1156	C 90	1120	Westchester	Sedan (fastback)	Plain	
810 2052	H	FB	1074			Convertible Phaeton			
810 2054	A					Westchester	Sedan (fastback)	Plain	
810 2057	H	FB	1077	C 91	137	Convertible Phaeton			
810 2061	H	FB	1252	C 91	134	Convertible Phaeton			
810 2062	A	FB	1303	C 90	1129	Westchester	Sedan (fastback)	Plain	
810 2066	A	FB	1055	C 90	1130	Westchester	Sedan (fastback)	Plain	
810 2069	A	FB	590	C 90	1042	Westchester	Sedan (fastback)	Pleated leather	Right hand drive.
810 2070	A	FB	1078	C 90	1117	Westchester	Sedan (fastback)	Plain	
810 2071	A	FB	1064	C 90	1125	Westchester	Sedan (fastback)	Plain	
810 2074	S	FB	1073	C 90	772	Beverly	Sedan (fastback)	Fixed armrests, front and rear	
810 2076	A			C 90	1140	Westchester	Sedan (fastback)	Plain	
810 2077	H	FB	1344			Convertible Phaeton			
810 2079	A	FB	1071	C 90	1187	Westchester	Sedan (fastback)	Plain	
810 2080	A	FB	1342			Westchester	Sedan (fastback)	Plain	
810 2081	A	FB	1339	C 90	1191	Westchester	Sedan (fastback)	Plain	
810 2086	A	FB	1333	C 90	1205	Westchester	Sedan (fastback)	Plain	
810 2087	A	FB	1338	C 90	1190	Westchester	Sedan (fastback)	Plain	
810 2093	A	FB	1326			Westchester	Sedan (fastback)	Plain	
810 2095	A	FB	1369	C 90	1311	Westchester	Sedan (fastback)	Plain	
810 2096	H					Convertible Phaeton			
810 2097	H	FB	2071			Convertible Phaeton			
810 2098	A					Westchester	Sedan (fastback)	Plain	
810 2099	H	FB	1045	C 91	142	Convertible Phaeton			
810		FB	922	C 91	149	Convertible Phaeton			
810 2105	A	FB	1318	C 90	1203	Westchester	Sedan (fastback)	Plain	
810 2106	A	FB	1347	C 90	1218	Westchester	Sedan (fastback)	Plain	
810 2112	A	FB	1363	C 90	1206	Westchester	Sedan (fastback)	Plain	Had higher front seat cushions.
810 2113	A	FB	1367	C 90	1210	Westchester	Sedan (fastback)	Plain	Had higher front seat cushions.
810 2114	A					Westchester	Sedan (fastback)	Plain	Had higher front seat cushions.
810 2115	H	FB	1867	C 91	150	Convertible Phaeton			
810 2116	A					Westchester	Sedan (fastback)	Plain	Had higher front seat cushions.
810 2117	A					Westchester	Sedan (fastback)	Plain	Had higher front seat cushions.
810 2118	A					Westchester	Sedan (fastback)	Plain	Had higher front seat cushions.
810 2121	A					Westchester	Sedan (fastback)	Plain	Had higher front seat cushions.
810 2122	A					Westchester	Sedan (fastback)	Plain	Had higher front seat cushions.
810 2124	A	FB	1340	C 90	1184	Westchester	Sedan (fastback)	Plain	Had higher front seat cushions, exported to Canada.
810 2125	H	FB	1316			Convertible Phaeton			
810 2126	A					Westchester	Sedan (fastback)	Plain	Had higher front seat cushions.
810 2127	A	FB	1330			Westchester	Sedan (fastback)	Plain	Exported to Canada.
810 2129	A	FB	1337	C 90	1213	Westchester	Sedan (fastback)	Plain	
810				C 90	1223	Westchester	Sedan (fastback)	Plain	
810 2132	A	FB	1319	C 90	1235	Westchester	Sedan (fastback)	Plain	
810 2134	A	FB	1372			Westchester	Sedan (fastback)	Plain	
810 2139	A	FB	1352	C 90	750	Westchester	Sedan (fastback)	Plain	
810 2140	A	FB	1356	C 90	751	Westchester	Sedan (fastback)	Plain	
810 2141	H	FB	1354	C 91	148	Convertible Phaeton			
810 2142	A	FB	1357			Westchester	Sedan (fastback)	Plain	
810 2144	A	FB	1385	C 90	1263	Westchester	Sedan (fastback)	Plain	

Serial #		Engine #		Body #			Model	Body Type	Interior pattern	Comments
810		FB	1390	C	90	1255	Westchester	Sedan (fastback)	Plain	
810	2147 H	FB	1206				Convertible Phaeton			
810	2148 H	FB	1388				Convertible Phaeton			Exported to Canada.
810	2150 H	FB	1577	C	91	160	Convertible Phaeton			
810	2150 A	FB	1417	C	90	1232	Westchester	Sedan (fastback)	Plain	
810	2151 H	FB	1421	C	91	173	Convertible Phaeton			
810	2152 A	FB	1419	C	90	1230	Westchester	Sedan (fastback)	Plain	Factory experimental car, for testing the supercharger.
810	2154 H	FB	1394	C	91	172	Convertible Phaeton			
810	2155 H	FB	1418	C	91	166	Convertible Phaeton			
810	2156 A	FB	1408	C	90	1179	Westchester	Sedan (fastback)	Plain	
810	2159 A	FB	1392	C	90	1239	Westchester	Sedan (fastback)	Plain	
810	2164 A	FB	1377				Westchester	Sedan (fastback)	Plain	
810	2165 A	FB	1248	C	90	1236	Westchester	Sedan (fastback)	Plain	
810	2167 H	FB	1397	C	91	167	Convertible Phaeton			
810	2169 H	FB	1410				Convertible Phaeton			
810	2170 H	FB	1378	C	91	178	Convertible Phaeton			
810	2172 A	FB	1404				Westchester	Sedan (fastback)	Plain	
810	2176 H	FB	1379	C	91	164	Convertible Phaeton			
810	2182 H	FB	1398	C	91	182	Convertible Phaeton			
810	2183 H	FB	1403	C	91	180	Convertible Phaeton			
810	2184 H	FB	1710				Convertible Phaeton			
810				C	91	184	Convertible Phaeton			
810	2186 H	FB	1264	C	91	183	Convertible Phaeton			
810	2187 H	FB	1293	C	91	181	Convertible Phaeton			
810	2188 A	FB	1432				Westchester	Sedan (fastback)	Plain	
810	2189 A						Westchester	Sedan (fastback)	Plain	
810	2190 A	FB	1414	C	90	1261	Westchester	Sedan (fastback)	Plain	
810	2191 H	FB	1430				Convertible Phaeton			
810	2192 A						Westchester	Sedan (fastback)	Plain	
810		FB	1435				Convertible Phaeton			
810		FB	1439				Convertible Phaeton			
810		FB	1458				Westchester	Sedan (fastback)	Plain	
810	2197 A	FB	1462				Westchester	Sedan (fastback)	Plain	
810	2198 H	FB	597				Convertible Phaeton			Right hand drive, exported to India.
810		FB	583				Convertible Phaeton			
810	2200 A	FB	1428	C	90	1268	Westchester	Sedan (fastback)	Plain	
810	2203 H	FB	1041				Convertible Phaeton			Right hand drive, exported to India.
810	2209 A	FB	1508	C	90	1221	Westchester	Sedan (fastback)	Plain	
810	2213 A	FB	1285	C	90	1283	Westchester	Sedan (fastback)	Plain	
810	2214 A	FB	1395				Westchester	Sedan (fastback)	Plain	
810	2216 H						Convertible Phaeton			
810	2220 A			C	90	1293	Westchester	Sedan (fastback)	Plain	
810	2221 H						Convertible Phaeton			
810	2224 H	FB	1063				Convertible Phaeton			
810	2226 H	FB	1444	C	91	190	Convertible Phaeton			
810	2228 H	FB	919				Convertible Phaeton			Right hand drive.
810	2232 H	FB	1505				Convertible Phaeton			
810	2233 H	FB	1516	C	91	204	Convertible Phaeton			
810	2238 A	FB	1519				Westchester	Sedan (fastback)	Plain	
810	2239 H						Convertible Phaeton			
810	2239 A	FB	1514	C	90	1292	Westchester	Sedan (fastback)	Plain	
810	2241 H	FB	1513				Convertible Phaeton			
810	2242 H	FB	1536	C	91	197	Convertible Phaeton			
810	2244 H	FB	1511				Convertible Phaeton			
810	2246 H	FB	1553	C	91	205	Convertible Phaeton			
810	2247 H	FB	1547	C	91	210	Convertible Phaeton			
810	2248 H						Convertible Phaeton			
810	2250 A	FB	1530				Westchester	Sedan (fastback)	Plain	Exported to Canada.
810	2253 H	FB	1442	C	91	207	Convertible Phaeton			
810	2254 H	FB	1556	C	91	322	Convertible Phaeton			
810		FB	1465				Convertible Phaeton			
810	2258 H						Convertible Phaeton			
810	2262 H	FB	1044				Convertible Phaeton			Right hand drive, exported to India.
810	2266 H			C	91	231	Convertible Phaeton			
810	2269 H			C	91	244	Convertible Phaeton			
810	2270 H	FB	1621	C	91	236	Convertible Phaeton			
810	2271 H	FB	1625	C	91	230	Convertible Phaeton			
810	2273 H	FB	1626				Convertible Phaeton			
810	2275 H	FB	1391	C	91	249	Convertible Phaeton			
810	2276 H	FB	1471	C	91	245	Convertible Phaeton			
810	2283 H	FB	1569	C	91	256	Convertible Phaeton			
810	2288 H	FB	1586	C	91	239	Convertible Phaeton			
810	2291 H	FB	592	C	91	198	Convertible Phaeton			Right hand drive.
810	2292 H	FB	1518				Convertible Phaeton			
810	2297 H	FB	942	C	91	260	Convertible Phaeton			
810	2298 H	FB	1534	C	91	255	Convertible Phaeton			
810	2299 H	FB	1515				Convertible Phaeton			
810		FB	1437	C	91	277	Convertible Phaeton			
810	2307 A	FB	1659	C	90	1290	Westchester	Sedan (fastback)	Plain	

Serial #	Engine #	Body #	Model	Body Type	Interior pattern	Comments
810	FB 1631		Westchester	Sedan (fastback)	Plain	
810 2311 H			Convertible Phaeton			
810 2312 H	FB 1637	C 91 232	Convertible Phaeton			
810 2316 H	FB 1520		Convertible Phaeton			
810 2317 A	FB 1501	C 90 1250	Westchester	Sedan (fastback)	Plain	
810 2320 A	FB 1493	C 90 1309	Westchester	Sedan (fastback)	Plain	
810 2330 H	FB 1449		Convertible Phaeton			
810	FB 1542		Convertible Phaeton			
810 2333 H	FB 1543	C 91 301	Convertible Phaeton			
810 2335 H	FB 1699	C 91 289	Convertible Phaeton			
810 2336 H	FB 1662	C 91 276	Convertible Phaeton			
810 2337 H		C 91 294	Convertible Phaeton			
810 2338 H	FB 283	C 91 184	Convertible Phaeton			
810 2339 H	FB 1640	C 91 302	Convertible Phaeton			
810	FB 538		Convertible Coupe			
810 2346 H	FB 1689		Convertible Phaeton			
810		C 92 115	Convertible Coupe			
810 2348 F	FB 1682	C 92 116	Convertible Coupe			
810	FB 1695		Convertible Coupe			
810 2350 H	FB 1541	C 91 293	Convertible Phaeton			
810 2352 H	FB 1450		Convertible Phaeton			
810 2357 H	FB 1692	C 91 241	Convertible Phaeton			
810 2360 F	FB 1567	C 92 125	Convertible Coupe			
810 2361 H	FB 1632	C 91 680	Convertible Phaeton			
810 2362 H	FB 1622	C 91 223	Convertible Phaeton			
810 2365 H	FB 1630	C 91 228	Convertible Phaeton			
810 2366 F	FB 310		Convertible Coupe			
810 2370 H	FB 1538	C 91 278	Convertible Phaeton			
810 2373 F	FB 1550	C 92 123	Convertible Coupe			
810 2374 H			Convertible Phaeton			
810 2375 F			Convertible Coupe			
810 2377 H	FB 1560	C 91 303	Convertible Phaeton			
810 2378 H	FB 1469		Convertible Phaeton			
810 2381 H	FB 1636		Convertible Phaeton			
810 2383 H	FB 1634		Convertible Phaeton			
810 2386 F	FB 1478	C 92 128	Convertible Coupe			
810 2387 F	FB 1506	C 92 131	Convertible Coupe			
810 2406 H	FB 1461		Convertible Phaeton			
810 2408 A	FB 1680	C 90 1189	Westchester	Sedan (fastback)	Pleated leather	
810 2410 H	FB 1596	C 91 317	Convertible Phaeton			
810 2411 H	FB 1529	C 91 323	Convertible Phaeton			
810 2412 H	FB 1537	C 91 328	Convertible Phaeton			
810 2413 H	FB 1657		Convertible Phaeton			
810 2417 H	FB 1643		Convertible Phaeton			
810 2418 H			Convertible Phaeton			
810 2419 H	FB 1544		Convertible Phaeton			
810 2420 H	FB 1561	C 91 353	Convertible Phaeton			
810 2423 H	FB 1638	C 91 343	Convertible Phaeton			
810 2424 H	FB 1608	C 91 344	Convertible Phaeton			
810 2426 H			Convertible Phaeton			
810		C 91 350	Convertible Phaeton			
810 2428 H	FB 1660	C 91 361	Convertible Phaeton			
810 2429 H		C 91 361	Convertible Phaeton			
810 2430 H	FB 1606	C 91 365	Convertible Phaeton			
810 2431 F		C 92 135	Convertible Coupe			
810 2432 H	FB 1654	C 91 325	Convertible Phaeton			
810		C 91 362	Convertible Phaeton			
810 2434 H	FB 1571	C 91 357	Convertible Phaeton			
810 2435 H	FB 1590		Convertible Phaeton			
810 2436 H	FB 1592	C 91 366	Convertible Phaeton			
810 2440 H	FB 1585	C 91 372	Convertible Phaeton			
810 2443 H	FB 1565		Convertible Phaeton			
810 2445 F	FB 1125	C 92 138	Convertible Coupe			Right hand drive.
810 2447 H	FB 1580	C 91 316	Convertible Phaeton			
810	FB 1708		Convertible Phaeton			
810 2450 H	FB 1574	C 91 377	Convertible Phaeton			
810 2452 H	FB 1800	C 91 380	Convertible Phaeton			
810 2455 F	FB 1724	C 92 147	Convertible Coupe			
810 2458 H	FB 1741		Convertible Phaeton			
810 2459 H	FB 1722		Convertible Phaeton			
810	FB 2088	C 91 315	Convertible Phaeton			
810 2462 H	FB 1716	C 91 344	Convertible Phaeton			
810 2466 H	FB 1703	C 91 374	Convertible Phaeton			
810 2468 F	FB 1765	C 92 142	Convertible Coupe			
810 2469 F	FB 1709	C 92 144	Convertible Coupe			
810 2471 H			Convertible Phaeton			
810 2473 F	FB 1674	C 92 155	Convertible Coupe			
810 2475 F	FB 1672	C 92 151	Convertible Coupe			
810 2477 A	FB 1712		Westchester	Sedan (fastback)	Plain	

Serial #		Engine #		Body #		Model	Body Type	Interior pattern	Comments
810 2481	A	FB	1673	C 90	1316	Westchester	Sedan (fastback)	Plain	
810 2482	F	FB	1715	C 92	152	Convertible Coupe			
810 2483	A	FB	1713			Westchester	Sedan (fastback)	Plain	
810 2485	H					Convertible Phaeton			
810 2487	F	FB	1736	C 92	153	Convertible Coupe			
810 2488	H	FB	1714	C 91	391	Convertible Phaeton			
810 2491	A	FB	948	C 90	1358	Westchester	Sedan (fastback)	Plain	Right hand drive.
810 2492	F	FB	1732	C 92	156	Convertible Coupe			
810 2496	F	FB	1768	C 92	160	Convertible Coupe			
810 2499	F	FB	929	C 92	165	Convertible Coupe			Right hand drive.
810 2500	F					Convertible Coupe			Right hand drive.
810				C 92	169				
810 2503	F					Convertible Coupe			
810 2505	F	FB	1737	C 92	171	Convertible Coupe			
810 2512	A	FB	1761	C 90	1323	Westchester	Sedan (fastback)	Plain	
810 2513	A	FB	1258			Westchester	Sedan (fastback)	Plain	
810 2515	F	FB	1774	C 92	181	Convertible Coupe			
810 2517	A	FB	1756			Westchester	Sedan (fastback)		
810 2518	F	FB	1764	C 92	178	Convertible Coupe			
810 2519	F	FB	1754	C 92	177	Convertible Coupe			
810 2520	F	FB	1735			Convertible Coupe			
810 2522	A	FB	1729	C 90	1302	Westchester	Sedan (fastback)	Plain	
810		FB	1760			Westchester	Sedan (fastback)	Plain	
810 2525	A	FB	1784			Westchester	Sedan (fastback)	Plain	
810 2526	A	FB	1847	C 90	1368	Westchester	Sedan (fastback)	Plain	
810 2530	A	FB	1759			Westchester	Sedan (fastback)	Plain	
810				C 92	189	Convertible Coupe			
810 2533	F	FB	1742	C 92	192	Convertible Coupe			
810 2535	F	FB	1763	C 92	187	Convertible Coupe			
810 2545	A	FB	1803			Westchester	Sedan (fastback)	Plain	
810 2547	A	FB	889	C 90	1330	Westchester	Sedan (fastback)	Plain	
810 2548	F	FB	1817	C 92	196	Convertible Coupe			
810 2554	A	FB	843			Westchester	Sedan (fastback)	Plain	
810 2556	A	FB	1667	C 90	1337	Westchester	Sedan (fastback)	Plain	
810 2563	A	FB	1032	C 90	1357	Westchester	Sedan (fastback)	Pleated leather	Right hand drive, exported to South Africa.
810 2569	A	FB	1849			Westchester	Sedan (fastback)	Plain	
810 2570	S	FB	3259	C 90	1319	Beverly	Sedan (fastback)	Fixed armrests, front and rear	May be A-C-D Company-installed engine.
810 2573	A	FB	505	C 90	1343	Westchester	Sedan (fastback)	Plain	
810 2579	A	FB	1880	C 90	1348	Westchester	Sedan (fastback)	Plain	
810 2581	A	FB	1900	C 90	1350	Westchester	Sedan (fastback)	Plain	
810 2591	F					Convertible Coupe			
810 2595	A			C 90	1519	Westchester	Sedan (fastback)	Plain	
810 2597	F	FB	1909	C 92	209	Convertible Coupe			
810 2598	H	FB	1902	C 91	411	Convertible Phaeton			
810 2606	H	FB	1781	C 91	392	Convertible Phaeton			
810 2610	F	FB	1808			Convertible Coupe			
810 2620	H	FB	1342			Convertible Phaeton			
810 2624	A	FB	1821	C 90	1354	Westchester	Sedan (fastback)	Plain	
810 2626	A					Westchester	Sedan (fastback)	Plain	
810 2638	A	FB	1601	C 90	1365	Westchester	Sedan (fastback)	Plain	
810 2641	A	FB	1482			Westchester	Sedan (fastback)	Plain	
810 2642	A					Westchester	Sedan (fastback)	Plain	
810 2648	A					Westchester	Sedan (fastback)	Plain	
810		FB	1910			Westchester	Sedan (fastback)	Plain	
810 2651	A	FB	1926	C 90	1375	Westchester	Sedan (fastback)	Pleated leather	May have had a rear-mounted spare.
810 2652	A					Westchester	Sedan (fastback)	Plain	
810 2661	F	FB	1968			Convertible Coupe			
810 2662	H					Convertible Phaeton			
810 2669	A	FB	1879			Westchester	Sedan (fastback)	Plain	
810 2670	A	FB	1869			Westchester	Sedan (fastback)	Plain	
810		FB	1961			Westchester	Sedan (fastback)	Plain	
810 2686	A	FB	1963	C 90	972	Westchester	Sedan (fastback)	Plain	
810 2690	H	FB	1949	C 91	386	Convertible Phaeton			
810 2695	A					Westchester	Sedan (fastback)	Plain	
810 2697	M	FB	1969	C 99	101	Hardtop Coupe			Created at the factory from a Convertible Coupe.
810 2698	A					Westchester	Sedan (fastback)	Plain	
810 2721	H	FB	2000	C 91	424	Convertible Phaeton			
810 2722	H	FB	2018	C 91	437	Convertible Phaeton			
810 2729	H	FB	1562			Convertible Phaeton			
810 2830	S	FB	1234			Beverly	Sedan (fastback)	Fixed armrests, front and rear	

253

1937 Model 812, Standard Wheelbase

The first several hundred cars sold as 812s were actually unsold 810s, renumbered. The first car built as an 812 was 812 1526A.

Serial #		Engine #		Body #		Model	Body Type	Interior Pattern	Comments
812 1029	A	FB	1311	C 90	123	Westchester	Sedan (fastback)	Plain	A handbuilt showcar, renumbered.
812		FB	1177	C 90	132	Westchester	Sedan (fastback)	Plain	A handbuilt showcar, renumbered.
812 1046	A	FB	1873	C 90	134	Westchester	Sedan (fastback)	Plain	A handbuilt showcar, renumbered.
812 1052	A	FB	1863	C 90	144	Westchester	Sedan (fastback)	Plain	A handbuilt showcar, renumbered.
812 1068	A	FB	2010	C 90	153	Westchester	Sedan (fastback)	Plain	A handbuilt showcar, renumbered. Later equipped with a bolt-on accessory trunk.
812 1103	A	FB	1298	C 90	1071	Westchester	Sedan (fastback)	Plain	
812 1104	A	FB	1853			Westchester	Sedan (fastback)	Plain	
812				C 92	173	Convertible Coupe			
812 31106	F	FC	2291	C 92	179	Convertible Coupe			The original engine was probably replaced with a supercharged one before this car was sold.
812 1110	F	FB	1677	C 92	134	Convertible Coupe			
812 1112	A	FB	1725	C 90	1317	Westchester	Sedan (fastback)	Plain	
812 1113	F	FB	106	C 92	158	Convertible Coupe			
812 1114	F			C 92	148	Convertible Coupe			
812 1115	A	FB	1345			Westchester	Sedan (fastback)	Plain	
812 1116	A	FB	1056			Westchester	Sedan (fastback)	Plain	
812 1117	A	FB	733	C 90	339	Westchester	Sedan (fastback)	Plain	
812 1118	H	FB	1628	C 91	246	Convertible Phaeton			
812		FB	1085			Convertible Phaeton			
812		FB	1460			Convertible Phaeton			
812 1126	F	FB	1481	C 92	167	Convertible Coupe			
812 1127	H	FB	1635	C 91	273	Convertible Phaeton			
812 1129	S	FB	1745			Beverly	Sedan (fastback)	Fixed armrests, front and rear	
812 1130	A	FB	1780	C 90	1298	Westchester	Sedan (fastback)	Plain	
812 1138	S	FB	723	C 90	576	Beverly	Sedan (fastback)	Fixed armrests, front and rear	
812 1139	S	FB	604			Beverly	Sedan (fastback)	Fixed armrests, front and rear	
812		FB	494			Westchester	Sedan (fastback)	Plain	Right hand drive.
812 1143	A	FB	954	C 90	726	Westchester	Sedan (fastback)	Plain	
812 1145	A	FB	1114	C 90	968	Westchester	Sedan (fastback)	Plain	
812		FB	1126			Convertible Phaeton			
812 1147	A	FB	1132			Westchester	Sedan (fastback)	Plain	
812 1151	A	FB	1384			Westchester	Sedan (fastback)	Plain	
812				C 91	158	Convertible Phaeton			
812		FB	1460			Convertible Phaeton			
812 1155	H	FB	1644	C 91	269	Convertible Phaeton			
812 1156	H	FB	1645			Convertible Phaeton			
812		FB	1665			Convertible Phaeton			
812 1159	H	FB	1486			Convertible Phaeton			
812 1160	H	FB	1611	C 91	233	Convertible Phaeton			
812 1161	H	FB	1555	C 91	274	Convertible Phaeton			
812 1162	H	FB	1559	C 91	284	Convertible Phaeton			
812 1163	F					Convertible Coupe			
812 1164	F	FB	1532	C 92	133	Convertible Coupe			
812 1165	H	FB	1686	C 91	333	Convertible Phaeton			
812 1167	F	FB	1577			Convertible Coupe			
812		FB	1585	C 91	350	Convertible Phaeton			
812 1170	A	FB	2115	C 90	1273	Westchester	Sedan (fastback)	Plain	
812 1173	S	FB	658	C 90	398	Beverly	Sedan (fastback)	Fixed armrests, front and rear	
812 1176	S	FB	662	C 90	437	Beverly	Sedan (fastback)	Fixed armrests, front and rear	
812 1177	H	FB	1578	C 91	317	Convertible Phaeton			
812		FB	1473			Convertible Phaeton			
812 1179	A	FB	1497	C 90	1222	Westchester	Sedan (fastback)	Plain	
812 1180	A	FB	1780			Westchester	Sedan (fastback)	Plain	
812 1186	H					Convertible Phaeton			
812 1187	A					Westchester	Sedan (fastback)	Plain	
812 1188	A					Westchester	Sedan (fastback)	Plain	
812		FB	1224			Convertible Phaeton			
812 1199	A	FB	651	C 90	558	Westchester	Sedan (fastback)	Plain	
812 1202	A	FB	1029	C 90	931	Westchester	Sedan (fastback)	Plain	
812 1203	A	FB	1358			Westchester	Sedan (fastback)	Plain	
812 1206	A	FB	1350	C 90	1238	Westchester	Sedan (fastback)	Plain	
812 1207	A	FB	1103	C 90	1008	Westchester	Sedan (fastback)	Plain	
812 1208	A	FB	1089			Westchester	Sedan (fastback)	Plain	
812 1209	S	FB	866			Beverly	Sedan (fastback)	Fixed armrests, front and rear	
812 1215	A	FB	1977	C 90	1372	Westchester	Sedan (fastback)	Plain	
812 1217	A	FB	1196	C 90	1064	Westchester	Sedan (fastback)	Plain	
812 1219	S	FB	775	C 90	691	Beverly	Sedan (fastback)	Fixed armrests, front and rear	
812 1221	S	FB	877	C 90	881	Beverly	Sedan (fastback)	Fixed armrests, front and rear	
812 1222	H	FB	1597	C 91	237	Convertible Phaeton			
812 1233	H	FB	1627	C 91	264	Convertible Phaeton			
812 1236	F	FB	1762	C 92	182	Convertible Coupe			
812 1238	H	FB	1770			Convertible Phaeton			
812 1239	F	FB	1827	C 92	225	Convertible Coupe			
812 1240	F	FB	1794	C 92	203	Convertible Coupe			

Serial #		Engine #		Body #		Model	Body Type	Interior Pattern	Comments
812	1244	A	FB	1758	C	90	1257	Westchester	Sedan (fastback)	Plain	
812	1248	S	FB	399	C	90	302	Beverly	Sedan(fastback)	Fixed armrests, front and rear	
812	1245	A	FB	1381				Westchester	Sedan (fastback)	Plain	
812	1254	A	FB	1409	C	90	1271	Westchester	Sedan (fastback)	Plain	
812	1256	F	FB	1459	C	92	129	Convertible Coupe			
812	1258	H	FB	1773	C	91	408	Convertible Phaeton			
812	1259	H	FB	521	C	91	340	Convertible Phaeton			
812	31262	A	FC	2122				Westchester	Sedan (fastback)	Plain	The original engine was probably replaced with a supercharged one before this car was sold.
812	1263	A	FB	639				Westchester	Sedan (fastback)	Plain	
812	1264	A	FB	1085	C	90	630	Westchester	Sedan (fastback)	Plain	
812	1267	A	FB	423				Westchester	Sedan (fastback)	Plain	
812	1269	F	FB	588				Convertible Coupe			
812	1270	H	FB	959	C	91	375	Convertible Phaeton			
812	1274	A	FB	1427				Westchester	Sedan (fastback)	Plain	
812	1275	A	FB	1983	C	90	1367	Westchester	Sedan (fastback)	Plain	
812	1277	A	FB	1916	C	90	1373	Westchester	Sedan (fastback)	Plain	
812	1281	A	FB	594	C	90	314	Westchester	Sedan(fastback)	Plain	
812	1291	A	FB	1209	C	90	1057	Westchester	Sedan (fastback)	Plain	
812	1292	A	FB	857	C	90	776	Westchester	Sedan (fastback)	Plain	
812	31293	A	FC	2128				Westchester	Sedan (fastback)	Plain	The original engine was probably replaced with a supercharged one before this car was sold.
812	1406	A	FB	1011	C	90	693	Westchester	Sedan (fastback)	Plain	
812	1409	A	FB	1151	C	90	1045	Westchester	Sedan (fastback)	Plain	
812	1413	A	FB	1174				Westchester	Sedan (fastback)	Plain	
812	1418	F	FB	1171	C	92	216	Convertible Coupe			
812	1419	A	FB	1180				Westchester	Sedan (fastback)	Plain	
812	1421	A	FB	1191				Westchester	Sedan (fastback)	Plain	
812	1422	A	FB	1193	C	90	962	Westchester	Sedan (fastback)	Plain	
812	1424	H						Convertible Phaeton			
812	1426	H	FB	1671	C	91	349	Convertible Phaeton			
812	1427	H	FB	1576				Convertible Phaeton			
812	1433	F	FB	1831				Convertible Coupe			
812	1435	F	FB	1753	C	92	221	Convertible Coupe			
812	1436	A	FB	522	C	90	1331	Westchester	Sedan (fastback)	Plain	
812	1437	F	FB	1483				Convertible Coupe			
812	1439	A						Westchester	Sedan (fastback)	Plain	
812	1443	F	FB	1783	C	92	180	Convertible Coupe			
812	1446	F	FB	1823				Convertible Coupe			
812	1448	H	FB	1834	C	91	407	Convertible Phaeton			
812	1449	H	FB	525	C	91	397	Convertible Phaeton			
812	1453	H	FB	1937	C	91	410	Convertible Phaeton			
812	1454	F	FB	1928				Convertible Coupe			
812	1455	F	FB	1881	C	92	243	Convertible Coupe			
812	1459	F	FB	1587	C	92	244	Convertible Coupe			
812	1460	H						Convertible Phaeton			
812	31462	F	FC	3118				Convertible Coupe			The original engine was probably replaced with a supercharged one before this car was sold.
812	1463	A			C	90	1370	Westchester	Sedan (fastback)	Plain	
812	1465	F	FB	1844				Convertible Coupe			
812	1466	A	FB	936	C	90	1288	Westchester	Sedan (fastback)	Plain	
812	31467	F	FC	3268	C	92	206	Convertible Coupe			The original engine was probably replaced with a supercharged one before this car was sold.
812	1468	S	FB	1820	C	96	101	Beverly	Sedan (bustleback)	Folding armrests, front and rear	The prototype bustleback Beverly, modified from an 810 sedan.
812					C	92	231	Convertible Coupe			
812	1469	H	FB	1495	C	91	393	Convertible Phaeton			
812	1471	A	FB	1960	C	90	1384	Westchester	Sedan (fastback)	Plain	
812	1472	H	FB	1979				Convertible Phaeton			
812	1474	H	FB	1850	C	91	395	Convertible Phaeton			
812	1477	H	FB	1893	C	91	399	Convertible Phaeton			
812	1479	F	FB	1886	C	92	222	Convertible Coupe			
812	1480	H	FB	1955				Convertible Phaeton			Had a rear-mounted spare.
812	1485	A	FB	1721				Westchester	Sedan (fastback)	Plain	
812	1487	S	FB	1981	C	90	786	Beverly	Sedan (fastback)	Fixed armrests, front and rear	
812	1490	F	FB	1842	C	98	101	Convertible Coupe			An experimental convertible with deeper frame members like the later Custom series. Modified from an 810 convertible coupe.
812											
812			FB	1855				Westchester	Sedan (fastback)	Plain	
812			FB	1973	C	91	433	Convertible Phaeton			
812	1493	H	FB	1965	C	91	423	Convertible Phaeton			
812	1494	H	FB	1966				Convertible Phaeton			
812	1495	A	FB	2016	C	90	1389	Westchester	Sedan (fastback)	Plain	
812	1497	A	FB	2002				Westchester	Sedan (fastback)	Plain	
812	1500	A	FB	2012	C	90	1395	Westchester	Sedan (fastback)	Pleated leather	Equipped with bolt-on accessory trunk.
812	1501	H	FB	2008	C	91	413	Convertible Phaeton			
812					C	90	1401	Westchester	Sedan (fastback)	Plain	
812	1507	H	FB	1967	C	91	419	Convertible Phaeton			
812	1509	A	FB	1996	C	90	1424	Westchester	Sedan (fastback)	Plain	
812	1510	A	FB	2004	C	90	1415	Westchester	Sedan (fastback)	Plain	

Serial #	Engine #			Body #			Model	Body Type	Interior Pattern	Comments
812 1511	A			C	90	1359	Westchester	Sedan (fastback)	Plain	Right hand drive
812 1513	H	FB	2017	C	91	421	Convertible Phaeton			
812 1518	S	FB	358				Beverly	Sedan (fastback)	Fixed armrests, front and rear	
812 1526	A	FB	2051	C	90	1419	Westchester	Sedan (fastback)	Plain	This was the first Cord built as an 812.
812				C	90	1426	Westchester	Sedan (fastback)	Plain	
812 1528	A	FB	2046				Westchester	Sedan (fastback)	Plain	Equipped with bolt-on accesory trunk.
812 1529	A	FB	2048				Westchester	Sedan (fastback)	Plain	
812 1531	H	FB	2379				Convertible Phaeton			
812 1531	A	FB	2043				Westchester	Sedan (fastback)	Plain	
812 1532	A	FB	2023	C	90	1036	Westchester	Sedan (fastback)	Plain	
812 1533	A	FB	2007	C	90	983	Westchester	Sedan (fastback)	Plain	
812 1536	A						Westchester	Sedan (fastback)	Plain	
812 1537	H	FB	2036	C	100	101	Convertible Phaeton			One of two phaetons with a bustle trunk.
812		FB	2041				Convertible Phaeton			
812 1541	A	FB	2034				Westchester	Sedan (fastback)	Plain	
812 1542	A	FB	2062				Westchester	Sedan (fastback)	Plain	
812 1545	H	FB	2033				Convertible Phaeton			
812 1546	H	FB	3258	C	91	425	Convertible Phaeton			May be A-C-D Company-installed engine.
812 1549	A	FB	2038	C	90	1405	Westchester	Sedan (fastback)	Plain	
812 1550	H	FB	2035				Convertible Phaeton			
812 1551	A	FB	1991				Westchester	Sedan (fastback)	Plain	
812 1553	H	FB	1997				Convertible Phaeton			
812 1554	H	FB	2024	C	91	435	Convertible Phaeton			
812		FB	2061				Beverly	Sedan (bustleback)	Folding armrests, front and rear	
812 1556	A	FB	2067	C	90	1038	Westchester	Sedan (fastback)	Plain	
812 1559	A	FB	2064	C	90	1402	Westchester	Sedan (fastback)	Plain	
812 1560	A	FB	2068				Westchester	Sedan (fastback)	Plain	
812 1567	A	FB	967				Westchester	Sedan (fastback)	Plain	
812 1570	A	FB	1324				Westchester	Sedan (fastback)	Plain	
812 1571	A	FB	2050				Westchester	Sedan (fastback)	Plain	
812				C	90	1401	Westchester	Sedan (fastback)	Plain	
812 1575	A	FB	1875	C	90	1406	Westchester	Sedan (fastback)	Plain	
812 1577	A	FB	1035				Westchester	Sedan (fastback)	Plain	Right hand drive, exported to South Africa.
812 1581	A	FB	1924				Westchester	Sedan (fastback)	Plain	
812 1583	A	FB	1990	C	90	1425	Westchester	Sedan (fastback)	Plain	
812 1586	A	FB	1915	C	90	1430	Westchester	Sedan (fastback)	Plain	
812 1587	A	FB	1862				Westchester	Sedan (fastback)	Plain	
812 1590	A	FB	1866	C	90	1439	Westchester	Sedan (fastback)	Plain	Equipped with bolt-on accesory trunk.
812				C	90	1448	Westchester	Sedan (fastback)	Plain	
812 1593	H	FB	1940				Convertible Phaeton			
812 1595	S			C	90	971	Beverly	Sedan (fastback)	Fixed armrests, front and rear	Right hand drive.
812 1596	A	FB	1950	C	90	1431	Westchester	Sedan (fastback)	Plain	
812 1597	A	FB	1953	C	90	967	Westchester	Sedan (fastback)	Plain	
812 1598	A	FB	1954				Westchester	Sedan (fastback)	Plain	
812 1599	S	FB	2589				Beverly	Sedan (fastback)	Fixed armrests, front and rear	May be A-C-D Company-installed engine.
812 1599	F	FB	1613	C	102	101	Convertible Coupe			Equipped with rumbleseat.
812 1600	K	FB	1987				Westchester	Sedan (fastback)	Plain	Exported to Australia. The serial suffix K is not used elsewhere. Right hand drive.
812 1601	K						Westchester	Sedan (fastback)	Plain	Exported to Australia. The serial suffix K is not used elsewhere. Right hand drive.
812 1602	M			C	99	102	Hardtop Coupe			Probably built for a customer who had seen 810 2697M.
812				C	90	1457	Westchester	Sedan (fastback)	Plain	
812 1603	A	FB	1084	C	90	1486	Westchester	Sedan (fastback)	Plain	
812				C	90	1460	Westchester	Sedan (fastback)	Plain	
812 1605	A	FB	815				Westchester	Sedan (fastback)	Plain	
812 1607	S	FB	2071	C	96	106	Beverly	Sedan (bustleback)	Folding armrests, front and rear	
812 1608	S	FB	2078	C	96	105	Beverly	Sedan (bustleback)	Folding armrests, front and rear	
812		FC	2520	C	91	358	Convertible Phaeton			
812				C	91	313	Convertible Phaeton			
812 31611	H	FC	2131				Convertible Phaeton			
812 31612	H	FC	2125	C	91	420	Convertible Phaeton			
812 31613	H	FC	2123	C	91	451	Convertible Phaeton			
812 31614	H	FC	2128				Convertible Phaeton			
812 1615	A	FB	1751				Westchester	Sedan (fastback)	Plain	
812 1616	S	FB	2152	C	96	109	Beverly	Sedan (bustleback)	Folding armrests, front and rear	
812 1617	S	FB	1858	C	96	112	Beverly	Sedan (bustleback)	Folding armrests, front and rear	
812 31618	H	FC	2126				Convertible Phaeton			
812 31619	H						Convertible Phaeton			
812 1620	A	FB	1750				Westchester	Sedan (fastback)	Plain	
812 1621	S	FB	1971	C	96	113	Beverly	Sedan (bustleback)	Folding armrests, front and rear	
812 1622	A	FB	1860	C	96	114	Westchester	Sedan (fastback)	Plain	Had a bustle trunk, like most Beverlys.
812 1623	S	FB	1749				Beverly	Sedan (bustleback)	Folding armrests, front and rear	
812 31624	H	FC	2127	C	91	458	Convertible Phaeton			
812 31625	H	FC	2132	C	91	449	Convertible Phaeton			
812 1626	S	FB	1672	C	96	116	Beverly	Sedan (bustleback)	Folding armrests, front and rear	
812 1627	S	FB	2381	C	96	117	Beverly	Sedan (bustleback)	Folding armrests, front and rear	
812 1629	A	FB	2083				Westchester	Sedan (fastback)	Plain	
812 1630	S	FB	1813	C	96	146	Beverly	Sedan (bustleback)	Folding armrests, front and rear	
812 1631	S	FB	2177				Beverly	Sedan (bustleback)	Folding armrests, front and rear	The engine with which this car was sold had originally been in a showcar.

Serial #		Engine #		Body #			Model	Body Type	Interior Pattern	Comments
812	1632	H						Convertible Phaeton		
812	1633	H	FB	1865	C	91	455	Convertible Phaeton		
812	1634	A						Westchester	Sedan (fastback)	Plain	The engine with which this car was sold had originally been in a showcar.
812	31635	S	FC	2134	C	96	145	Beverly	Sedan (bustleback)	Folding armrests, front and rear	
812	1636	S	FB	2361	C	96	118	Beverly	Sedan (bustleback)	Folding armrests, front and rear	
812	1637	S						Beverly	Sedan (bustleback)	Folding armrests, front and rear	The engine with which this car was sold had originally been in a showcar.
812	1638	A						Westchester	Sedan (fastback)	Plain	The engine with which this car was sold had originally been in a showcar.
812	1642	H	FB	640				Convertible Phaeton		
812	31643	H	FC	2133	C	91	439	Convertible Phaeton		
812	1645	A	FB	2184	C	90	1497	Westchester	Sedan (fastback)	Plain	The engine with which this car was sold had originally been in a showcar.
812	31646	H	FC	2136	C	91	463	Convertible Phaeton			Exported to Canada.
812	31648	H						Convertible Phaeton		
812	31650	S	FC	2141				Beverly	Sedan (bustleback)	Folding armrests, front and rear	
812	31651	H	FC	2145				Convertible Phaeton		
812	31652	H	FC	2143	C	91	469	Convertible Phaeton		
812			FC	2155				Convertible Phaeton		
812	31653	F	FC	2146	C	92	267	Convertible Coupe		
812	31654	H	FC	2267	C	91	464	Convertible Phaeton			Exported to Canada.
812	31655	F	FC	2260	C	92	266	Convertible Coupe		
812	31657	H	FC	2140	C	91	475	Convertible Phaeton		
812	1661	S	FB	1917	C	96	134	Beverly	Sedan (bustleback)	Folding armrests, front and rear	
812	1662	S	FB	1811				Beverly	Sedan (bustleback)	Folding armrests, front and rear	
812	1666	A	FB	1903				Westchester	Sedan (fastback)	Plain	
812	1668	S	FB	1947				Beverly	Sedan (bustleback)	Folding armrests, front and rear	
812	1669	H	FB	2101	C	91	467	Convertible Phaeton		
812	1674	A	FB	819	C	90	1503	Westchester	Sedan (fastback)	Plain	
812	31674	H						Convertible Phaeton		
812	1675	H	FB	2114	C	91	461	Convertible Phaeton		
812	1677	S	FB	2111				Beverly	Sedan (bustleback)	Folding armrests, front and rear	
812	1678	A	FB	2107	C	96	182	Westchester	Sedan (bustleback)	Plain	Had a bustle trunk, like most Beverlys.
812	1681	S	FB	2109	C	96	131	Beverly	Sedan (bustleback)	Folding armrests, front and rear	
812	1683	A	FB	2073	C	90	1506	Westchester	Sedan (fastback)	Plain	
812	1686	F	FB	2100	C	92	272	Convertible Coupe		
812	31690	H	FC	2265				Convertible Phaeton		
812	31693	F	FC	2277	C	92	271	Convertible Coupe		
812	31694	F	FC	2279	C	92	273	Convertible Coupe			Right hand drive.
812	31695	H	FC	2278	C	91	426	Convertible Phaeton		
812	31698	H	FC	2286	C	91	473	Convertible Phaeton		
812	31700	H	FC	2287	C	91	445	Convertible Phaeton		
812	31701	H	FC	2289	C	91	443	Convertible Phaeton		
812	31702	A	FC	2276				Westchester	Sedan (fastback)	Plain	Right hand drive.
812	1704	A	FB	2105				Westchester	Sedan (fastback)	Plain	
812	1705	A	FB	907	C	90	1463	Westchester	Sedan (fastback)	Plain	
812	31707	H	FC	2290				Convertible Phaeton		
812	31709	A	FC	2295	C	90	1461	Westchester	Sedan (fastback)	Plain	
812	31710	H	FC	2263	C	91	478	Convertible Phaeton		
812	31711	H	FC	2144				Convertible Phaeton		
812	31712	S	FC	2273				Beverly	Sedan (bustleback)	Folding armrests, front and rear	
812	1713	A	FB	2086				Westchester	Sedan (fastback)	Plain	
812	1715	S	FB	2384				Beverly	Sedan (bustleback)	Folding armrests, front and rear	
812	1719	S	FB	2103	C	96	240	Beverly	Sedan (bustleback)	Folding armrests, front and rear	
812	31722	H	FC	2274	C	91	444	Convertible Phaeton		
812	1725	H	FB	2104	C	91	447	Convertible Phaeton		
812					C	91	449	Convertible Phaeton		
812	31727	S	FC	2307	C	96	121	Beverly	Sedan (bustleback)	Folding armrests, front and rear	
812	31730	S	FC	2296	C	96	154	Beverly	Sedan (bustleback)	Folding armrests, front and rear	
812					C	96	153	Beverly	Sedan (bustleback)	Folding armrests, front and rear	
812	1733	H	FB	2099	C	91	482	Convertible Phaeton		
812	31735	F			C	92	269	Convertible Coupe		
812	31736	A	FC	2298	C	96	156	Westchester	Sedan (bustleback)	Plain	Had a bustle trunk, like most Beverlys.
812	31737	H	FC	2300	C	91	477	Convertible Phaeton		
812	31737	S	FC	2268				Beverly	Sedan (bustleback)	Folding armrests, front and rear	
812	31738	F	FC	2299	C	92	274	Convertible Coupe			Right hand drive.
812	1739	S			C	96	137	Beverly	Sedan (bustleback)	Folding armrests, front and rear	
812	1742	H	FB	2043				Convertible Phaeton		
812	1743	A	FB	2154	C	90	1478	Westchester	Sedan (fastback)	Plain	
812	1745	A	FB	2155	C	96	178	Westchester	Sedan (bustleback)	Plain	Had a bustle trunk, like most Beverlys.
812	1746	A	FB	2150	C	90	1449	Westchester	Sedan (fastback)	Plain	
812	31747	S	FC	2304	C	96	183	Beverly	Sedan (bustleback)	Folding armrests, front and rear	
812	1748	S	FB	2354	C	96	135	Beverly	Sedan (bustleback)	Folding armrests, front and rear	
812			FB	2180				Beverly	Sedan (bustleback)	Folding armrests, front and rear	
812	31750	A	FC	2313	C	90	1432	Westchester	Sedan (fastback)	Plain	
812	1751	H	FB	2453				Convertible Phaeton		
812	1752	A	FB	2271				Westchester	Sedan (bustleback)	Plain	Had a bustle trunk, like most Beverlys.
812	31753	H	FC	2315	C	91	489	Convertible Phaeton		
812	1754	H	FB	2168	C	91	485	Convertible Phaeton		

Serial #		Engine #		Body #			Model	Body Type	Interior Pattern	Comments
812		FB	2195				Westchester	Sedan (fastback)	Plain	
812 1764	A	FB	2221	C	96	163	Westchester	Sedan (bustleback)	Plain	Had a bustle trunk, like most Beverlys.
812 1765	H	FB	2227	C	91	476	Convertible Phaeton			
812 1766	H	FB	2357	C	91	475	Convertible Phaeton			
812 1769	F	FB	2352	C	92	246	Convertible Coupe			
812 1771	A	FB	2358	C	90	990	Westchester	Sedan (fastback)	Plain	
812 1772	A	FB	2356				Westchester	Sedan (fastback)	Plain	
812 31779	H	FC	2403	C	91	481	Convertible Phaeton			
812 31780	F	FC	2401	C	92	257	Convertible Coupe			
812 31781	A	FC	2408	C	90	1477	Westchester	Sedan (fastback)	Plain	
812 1782	H						Convertible Phaeton			
812 1783	H	FB	2337				Convertible Phaeton			
812 1784	S	FB	2415	C	96	159	Beverly	Sedan (bustleback)	Folding armrests, front and rear	
812 1785	H	FB	2331	C	91	486	Convertible Phaeton			
812		FB	2328				Beverly	Sedan (bustleback)	Folding armrests, front and rear	
812 1789	S	FB	2326	C	96	201	Beverly	Sedan (bustleback)	Folding armrests, front and rear	
812 1790	S	FB	2324	C	96	174	Beverly	Sedan (bustleback)	Folding armrests, front and rear	
812 1792	H	FB	2353				Convertible Phaeton			
812 1793	S	FB	2376	C	96	202	Beverly	Sedan (bustleback)	Folding armrests, front and rear	
812 1794	H	FB	2370	C	91	500	Convertible Phaeton			
812 1795	S	FB	2368	C	96	203	Beverly	Sedan (bustleback)	Folding armrests, front and rear	
812 1796	S	FB	2365				Beverly	Sedan (bustleback)	Folding armrests, front and rear	
812 1797	A	FB	2369				Westchester	Sedan (fastback)	Plain	
812 1798	S	FB	2363	C	96	190	Beverly	Sedan (bustleback)	Folding armrests, front and rear	
812 1799	H	FB	2360	C	91	499	Convertible Phaeton			
812 1800	S	FB	2378	C	96	169	Beverly	Sedan (bustleback)	Folding armrests, front and rear	
812 31801	H	FC	2571				Convertible Phaeton			
812 31804	A	FC	2513	C	96	122	Westchester	Sedan (bustleback)	Plain	Had a bustle trunk, like most Beverlys.
812 1806	A						Westchester	Sedan (fastback)	Plain	
812 1807	H	FB	2374				Convertible Phaeton			
812				C	91	501	Convertible Phaeton			
812 1814	H	FB	2372	C	91	504	Convertible Phaeton			
812		FB	2432				Westchester	Sedan (fastback)	Plain	
812 1818	S						Beverly	Sedan (bustleback)	Folding armrests, front and rear	
812 1819	F	FB	2424	C	92	235	Convertible Coupe			
812 1820	H	FB	2459				Convertible Phaeton			
812 1821	H	FB	2457	C	91	487	Convertible Phaeton			
812 1822	S	FB	2455				Beverly	Sedan (bustleback)	Folding armrests, front and rear	
812 1823	A	FB	2377	C	90	953	Westchester	Sedan (fastback)	Plain	
812 1825	S	FB	2438	C	96	193	Beverly	Sedan (bustleback)	Folding armrests, front and rear	
812 1827	H	FB	2447				Convertible Phaeton			Had a rear-mounted spare.
812		FB	2454				Convertible Phaeton			
812 1832	S	FB	2445				Beverly	Sedan (bustleback)	Folding armrests, front and rear	
812 1833	H	FB	2465				Convertible Phaeton			
812 1834	S	FB	2440	C	96	176	Beverly	Sedan (bustleback)	Folding armrests, front and rear	
812 31835	F	FC	2520				Convertible Coupe			
812 1835	F			C	92	262	Convertible Coupe			
812 31837	H	FC	2515	C	91	509	Convertible Phaeton			
812 31838	H	FC	2517	C	91	506	Convertible Phaeton			
812 31839	S	FC	2525	C	96	184	Beverly	Sedan (bustleback)	Folding armrests, front and rear	
812 31840	H	FC	2518	C	91	510	Convertible Phaeton			
812 31841	A	FC	2523	C	90	1452	Westchester	Sedan (fastback)	Plain	
812 31842	A	FC	3318	C	90	1469	Westchester	Sedan (fastback)	Plain	May be A-C-D Company-installed engine.
812 31843	H	FC	2519				Convertible Phaeton			
812 31845	S			C	96	186	Beverly	Sedan (bustleback)	Folding armrests, front and rear	
812 31846	S	FC	2536	C	96	196	Beverly	Sedan (bustleback)	Folding armrests, front and rear	
812		FC	2549				Beverly	Sedan (bustleback)	Folding armrests, front and rear	
812 31848	H	FC	2531				Convertible Phaeton			
812 1850	S	FB	2468	C	96	205	Beverly	Sedan (bustleback)	Folding armrests, front and rear	
812 31851	H	FC	2530	C	91	512	Convertible Phaeton			
812 1852	S	FB	2467	C	96	170	Beverly	Sedan (bustleback)	Folding armrests, front and rear	
812 31853	H	FC	2534	C	91	494	Convertible Phaeton			Right hand drive.
812 31854	A	FC	2528	C	90	1409	Westchester	Sedan (fastback)	Plain	
812 31855	A	FC	2529	C	96	210	Westchester	Sedan (fastback)	Plain	Had a bustle trunk, like most Beverlys.
812 1856	H	FB	2383	C	91	513	Convertible Phaeton			
812 31859	A	FC	2521				Westchester	Sedan (fastback)	Plain	
812		FC	2538	C	96	971	Beverly	Sedan (bustleback)	Folding armrests, front and rear	
812 1861	S	FB	2472	C	96	178	Beverly	Sedan (bustleback)	Folding armrests, front and rear	
812 1862	H	FB	2473				Convertible Phaeton			
812 1864	S	FB	2429				Beverly	Sedan (bustleback)	Folding armrests, front and rear	
812 31865	H	FC	2524	C	91	490	Convertible Phaeton			
812 31866	H						Convertible Phaeton			
812 31867	F	FC	2572	C	92	237	Convertible Coupe			
812		FC	2595	C	91	412	Convertible Phaeton			
812 31868	H	FC	2596	C	91	518	Convertible Phaeton			
812 31869	A			C	90	405	Westchester	Sedan (fastback)	Plain	
812 1872	H	FB	2385	C	91	519	Convertible Phaeton			
812 31873	A	FC	2143	C	90	1525	Westchester	Sedan (fastback)	Pleated leather	
812 31874	A	FC	2595	C	90	1514	Westchester	Sedan (fastback)	Plain	

Serial #		Engine #		Body #			Model	Body Type	Interior Pattern	Comments
812 1875	H	FB	2388				Convertible Phaeton			
812 1876	H	FB	2380	C	91	523	Convertible Phaeton			
812 31879	A	FC	2567				Westchester	Sedan (fastback)	Plain	
812		FC	2569				Westchester	Sedan (fastback)	Plain	
812				C	92	231	Convertible Coupe			
812 31882	F	FC	2568	C	92	245	Convertible Coupe			
812 1887	H	FB	2546	C	91	527	Convertible Phaeton			
812				C	91	531	Convertible Phaeton			
812 1889	H	FB	2456	C	91	532	Convertible Phaeton			
812 31891	H	FC	2571	C	91	521	Convertible Phaeton			
812 1892	H	FB	2466	C	91	533	Convertible Phaeton			
812 1893	H	FB	2396				Convertible Phaeton			
812 31895	S	FC	2638	C	96	237	Beverly	Sedan (bustleback)	Folding armrests, front and rear	
812 31905	H	FC	2631				Convertible Phaeton			
812 1907	H	FB	2474				Convertible Phaeton			
812 31909	S	FC	2632	C	96	242	Beverly	Sedan (bustleback)	Folding armrests, front and rear	
812 31910	H	FC	2634				Convertible Phaeton			This was Tom Mix' car. It had a rear-mounted spare.
812 31911	S	FC	2639				Beverly	Sedan (bustleback)	Folding armrests, front and rear	
812 31914	H	FC	2615	C	91	536	Convertible Phaeton			
812 31915	H	FC	2735	C	91	538	Convertible Phaeton			
812 31916	S	FC	2622	C	96	235	Beverly	Sedan (bustleback)	Folding armrests, front and rear	
812 31917	H	FC	2633				Convertible Phaeton			
812 31918	H	FC	2647	C	91	540	Convertible Phaeton			
812 31919	S	FC	2623	C	96	247	Beverly	Sedan (bustleback)	Folding armrests, front and rear	
812 31920	S						Beverly	Sedan (bustleback)	Folding armrests, front and rear	
812 31921	S	FC	2660				Beverly	Sedan (bustleback)	Folding armrests, front and rear	
812 1922	F	FB	2475	C	92	256	Convertible Coupe			
812 1924	A						Westchester	Sedan (fastback)	Plain	
812 1926	A			C	90	533	Westchester	Sedan (fastback)	Plain	Right hand drive
812 31927	H	FC	2617	C	91	547	Convertible Phaeton			
812 31929	A	FC	2627				Westchester	Sedan (fastback)	Plain	
812 1930	H	FB	2581	C	100	102	Convertible Phaeton			One of two phaetons with a bustle trunk.
812		FB	2579				Beverly	Sedan (bustleback)	Folding armrests, front and rear	
812 1932	H	FB	2580	C	91	545	Convertible Phaeton			
812 1933	H	FB	2578	C	91	549	Convertible Phaeton			
812 31934	S	FC	2628	C	96	232	Beverly	Sedan (bustleback)	Folding armrests, front and rear	
812 1935	S						Beverly	Sedan (bustleback)	Folding armrests, front and rear	
812 1936	H	FB	2586				Convertible Phaeton			Exported to Canada.
812		FB	2604	C	90	1517	Westchester	Sedan (fastback)	Plain	Right hand drive.
812 31937	S	FC	2649	C	96	265	Beverly	Sedan (bustleback)	Folding armrests, front and rear	
812 31937	F	FC	2618	C	92	242	Convertible Coupe			
812 1939	S	FB	2478				Beverly	Sedan (bustleback)	Folding armrests, front and rear	
812 31940	H	FC	2752				Convertible Phaeton			
812 31941	A	FC	2658	C	90	1531	Westchester	Sedan (fastback)	Plain	
812 31942	A	FC	2723	C	90	1532	Westchester	Sedan (fastback)	Plain	
812 1943	S	FB	2480	C	96	241	Beverly	Sedan (bustleback)	Folding armrests, front and rear	
812 31945	F	FC	2705	C	92	251	Convertible Coupe			
812 31946	H	FC	2669	C	91	563	Convertible Phaeton			
812 31948	F	FC	2708				Convertible Coupe			
812				C	96	246	Beverly	Sedan (bustleback)	Folding armrests, front and rear	
812 31950	F	FC	3008	C	92	249	Convertible Coupe			
812 31955	A	FC	2704	C	90	1521	Westchester	Sedan (fastback)	Plain	
812 31959	S	FC	2692	C	96	240	Beverly	Sedan (bustleback)	Folding armrests, front and rear	
812 31960	A	FC	2693	C	90	1515	Westchester	Sedan (fastback)	Plain	
812 31961	H	FC	2695	C	91	558	Convertible Phaeton			
812 31962	F	FC	2696				Convertible Coupe			
812				C	91	578	Convertible Phaeton			
812 31964	S	FC	2697	C	96	256	Beverly	Sedan (bustleback)	Folding armrests, front and rear	
812 31966	H	FC	2701	C	91	579	Convertible Phaeton			
812		FB	2738				Convertible Phaeton			
812		FC	2729				Beverly	Sedan (bustleback)	Folding armrests, front and rear	
812 31970	S	FC	2730				Beverly	Sedan (bustleback)	Folding armrests, front and rear	
812 31972	S	FC	2702	C	96	227	Beverly	Sedan (bustleback)	Folding armrests, front and rear	
812 31976	S	FC	2674				Beverly	Sedan (bustleback)	Folding armrests, front and rear	
812 31977	S	FC	2673	C	96	215	Beverly	Sedan (bustleback)	Folding armrests, front and rear	
812 1979	S			C	96	249	Beverly	Sedan (bustleback)	Folding armrests, front and rear	
812 1990	A	FB	2469	C	96	218	Westchester	Sedan (bustleback)	Plain	Had a bustle trunk, like most Beverlys.
812 1991	S	FB	2501				Beverly	Sedan (bustleback)	Folding armrests, front and rear	
812 1993	H						Convertible Phaeton			
812 1994	S	FB	2495	C	96	286	Beverly	Sedan (bustleback)	Folding armrests, front and rear	
812 31996	S	FC	2792	C	96	236	Beverly	Sedan (bustleback)	Folding armrests, front and rear	
812 32000	H	FC	2653	C	91	588	Convertible Phaeton			
812 2001	S	FB	2585	C	96	281	Beverly	Sedan (bustleback)	Folding armrests, front and rear	
812 32003	A	FC	2689	C	90	1519	Westchester	Sedan (fastback)	Plain	
812 2005	A	FB	2564	C	90	891	Westchester	Sedan (fastback)	Plain	
812		FC	2726				Convertible Phaeton			
812		FC	2729				Convertible Phaeton			
812		FB	2750				Westchester	Sedan (fastback)	Plain	
812 32007	F	FC	2682	C	92	250	Convertible Coupe			

Serial #		Engine #		Body #			Model	Body Type	Interior Pattern	Comments
812 32008	S	FC	2887				Beverly	Sedan (bustleback)	Folding armrests, front and rear	
812 32011	H	FC	2767	C	91	572	Convertible Phaeton			
812 32012	A	FC	2768	C	90	961	Westchester	Sedan (fastback)	Plain	
812 32013	H	FC	2773	C	91	588	Convertible Phaeton			
812 32014	S	FC	2775				Beverly	Sedan (bustleback)	Folding armrests, front and rear	
812 32016	A	FC	2774	C	90	1522	Westchester	Sedan (fastback)	Plain	
812 32017	H	FC	2766				Convertible Phaeton			
812 32021	S	FC	2794	C	96	224	Beverly	Sedan (bustleback)	Folding armrests, front and rear	
812 32023	F	FC	2783				Convertible Coupe			
812 32025	S	FC	2780				Beverly	Sedan (bustleback)	Folding armrests, front and rear	
812 32026	S	FC	2771	C	96	307	Beverly	Sedan (bustleback)	Folding armrests, front and rear	
812 2027	A	FB	2743	C	90	1560	Westchester	Sedan (fastback)	Plain	
812		FB	2750				Westchester	Sedan (fastback)	Plain	
812 2029	S	FB	2836	C	96	278	Beverly	Sedan (bustleback)	Folding armrests, front and rear	
812 2030	S	FB	2835	C	96	296	Beverly	Sedan (bustleback)	Folding armrests, front and rear	
812 2031	A	FB	2833	C	90	957	Westchester	Sedan (fastback)	Plain	
812 32032	S	FC	2765	C	96	291	Beverly	Sedan (bustleback)	Folding armrests, front and rear	
812 2035	S	FB	2099				Beverly	Sedan (bustleback)	Folding armrests, front and rear	
812		FC	2740				Beverly	Sedan (bustleback)	Folding armrests, front and rear	
812 2036	F			C	92	261	Convertible Coupe			
812 32037	S	FC	2745	C	96	303	Beverly	Sedan (bustleback)	Folding armrests, front and rear	
812 32038	S			C	96	283	Beverly	Sedan (bustleback)	Folding armrests, front and rear	This car has an eight-stack exhaust system.
812 2042	S	FB	2865				Beverly	Sedan (bustleback)	Folding armrests, front and rear	
812 2045	F			C	92	127	Convertible Coupe			
812 32046	H	FC	2796	C	91	575	Convertible Phaeton			
812 2048	A						Westchester	Sedan (fastback)	Plain	
812 2049	A	FB	2864				Westchester	Sedan (fastback)	Plain	
812 32050	S	FC	2799				Beverly	Sedan (bustleback)	Folding armrests, front and rear	
812 32051	S	FC	2806				Beverly	Sedan (bustleback)	Folding armrests, front and rear	
812 32052	S	FC	2790	C	96	253	Beverly	Sedan (bustleback)	Folding armrests, front and rear	
812 2056	S	FB	2391	C	96	298	Beverly	Sedan (bustleback)	Folding armrests, front and rear	
812 2057	S	FB	2611	C	96	264	Beverly	Sedan (bustleback)	Folding armrests, front and rear	
812 32058	A	FC	2808				Westchester	Sedan (fastback)	Plain	
812 32059	S	FC	2784	C	96	272	Beverly	Sedan (bustleback)	Folding armrests, front and rear	
812 2063	S	FB	2600	C	96	365	Beverly	Sedan (bustleback)	Folding armrests, front and rear	
812 32064	S	FC	2797				Beverly	Sedan (bustleback)	Folding armrests, front and rear	
812 2065	H	FB	2866				Convertible Phaeton			
812 32067	H	FC	2798				Convertible Phaeton			
812 32073	S	FC	2728	C	96	297	Beverly	Sedan (bustleback)	Folding armrests, front and rear	
812 2075	S	FB	2849				Beverly	Sedan (bustleback)	Folding armrests, front and rear	
812 32077	H	FC	2805				Convertible Phaeton			Had a rear-mounted spare.
812 32078	H	FC	2793	C	91	586	Convertible Phaeton			
812 2082	H	FB	1446				Convertible Phaeton			
812 32082	S	FC	2722	C	96	363	Beverly	Sedan (bustleback)	Folding armrests, front and rear	
812		FC	2822				Beverly	Sedan (bustleback)	Folding armrests, front and rear	
812 2086	S	FB	2862				Beverly	Sedan (bustleback)	Folding armrests, front and rear	
812 32089	S	FC	2873				Beverly	Sedan (bustleback)	Folding armrests, front and rear	
812 32090	S	FC	2885	C	96	367	Beverly	Sedan (bustleback)	Folding armrests, front and rear	
812 2093	S	FB	2843	C	96	260	Beverly	Sedan (bustleback)	Folding armrests, front and rear	
812 2095	S	FB	2844				Beverly	Sedan (bustleback)	Folding armrests, front and rear	
812 32096	S	FC	2889	C	96	377	Beverly	Sedan (bustleback)	Folding armrests, front and rear	
812 32097	F	FC	2884	C	92	176	Convertible Coupe			
812 2098	S	FB	2851				Beverly	Sedan (bustleback)	Folding armrests, front and rear	
812 32104	H	FC	2886	C	91	586	Convertible Phaeton			
812 32111	S	FC	2811				Beverly	Sedan (bustleback)	Folding armrests, front and rear	
812 32112	S	FC	2881				Beverly	Sedan (bustleback)	Folding armrests, front and rear	
812		FC	2882	C	91	552	Convertible Phaeton			
812 32116	H	FC	2888	C	91	541	Convertible Phaeton			
812 2117	S	FB	2858	C	96	353	Beverly	Sedan (bustleback)	Folding armrests, front and rear	
812 32119	A	FC	2825	C	90	1530	Westchester	Sedan (fastback)	Plain	
812 32121	H	FC	2813				Convertible Phaeton			
812 32123	S	FC	2911				Beverly	Sedan (bustleback)	Folding armrests, front and rear	
812 32125	S	FC	2902				Beverly	Sedan (bustleback)	Folding armrests, front and rear	
812 32127	H	FC	2907	C	91	574	Convertible Phaeton			Had a rear-mounted spare.
812 32128	F	FC	3047	C	92	240	Convertible Coupe			
812 32129	H	FC	2892	C	91	570	Convertible Phaeton			
812 32132	F	FC	2895	C	92	280	Convertible Coupe			
812				C	92	281	Convertible Coupe			
812 2134	S						Beverly	Sedan (bustleback)	Folding armrests, front and rear	
812 32135	S	FC	2898	C	96	326	Beverly	Sedan (bustleback)	Folding armrests, front and rear	
812 32138	S	FC	2908				Beverly	Sedan (bustleback)	Folding armrests, front and rear	
812 32140	S	FC	2893	C	96	175	Beverly	Sedan (bustleback)	Folding armrests, front and rear	This was Car #1 in the Stevens Trophy record run in 1937.
812 2141	H	FB	2718				Convertible Phaeton			Right hand drive.
812 2142	S	FB	2868				Beverly	Sedan (bustleback)	Folding armrests, front and rear	
812 32144	H			C	91	555	Convertible Phaeton			
812		FC	2921				Beverly	Sedan (bustleback)	Folding armrests, front and rear	
812 32148	S	FC	2965				Beverly	Sedan (bustleback)	Folding armrests, front and rear	
812 32150	S	FC	2947				Beverly	Sedan (bustleback)	Folding armrests, front and rear	

Serial #		Engine #		Body #			Model	Body Type	Interior Pattern	Comments
812 32151	F	FC	2948				Convertible Coupe			Right hand drive.
812 2152	S	FB	2605				Beverly	Sedan (bustleback)	Folding armrests, front and rear	
812 2153	S	FB	2606	C	96	314	Beverly	Sedan (bustleback)	Folding armrests, front and rear	
812				C	96	316	Beverly	Sedan (bustleback)	Folding armrests, front and rear	
812				C	96	358	Beverly	Sedan (bustleback)	Folding armrests, front and rear	
812 32155	S	FC	2937	C	96	341	Beverly	Sedan (bustleback)	Folding armrests, front and rear	
812 2156	H	FB	2742	C	91	559	Convertible Phaeton			Dual side-mounted spare tires.
812 32157	S	FC	2936	C	96	320	Beverly	Sedan (bustleback)	Folding armrests, front and rear	
812 2162	S	FB	2158				Beverly	Sedan (bustleback)	Folding armrests, front and rear	
812 32163	F	FC	2821				Convertible Coupe			
812 32166	H	FC	2917	C	91	556	Convertible Phaeton			
812 32169	H	FC	2929	C	91	589	Convertible Phaeton			
812 32170	H	FC	2930	C	91	129	Convertible Phaeton			
812 2173	H	FB	2749	C	91	594	Convertible Phaeton			
812 32179	H	FC	2968	C	91	597	Convertible Phaeton			
812 32180	H	FC	2969	C	91	576	Convertible Phaeton			
812 32181	A	FC	2970				Westchester	Sedan (fastback)	Plain	
812 32184	F	FC	2903	C	92	283	Convertible Coupe			Right hand drive, exported to South Africa.
812 32185	H	FC	2960	C	91	582	Convertible Phaeton			
812 32186	H						Convertible Phaeton			
812 32187	H	FC	2963	C	91	573	Convertible Phaeton			
812 2190	F	FB	2870	C	92	278	Convertible Coupe			
812 32191	A						Westchester	Sedan (fastback)	Plain	
812 32194	H	FC	3303	C	91	596	Convertible Phaeton			May be A-C-D Company-installed engine.
812 2195	S	FB	2199				Beverly	Sedan (bustleback)	Folding armrests, front and rear	
812 32198	H	FC	2975	C	91	583	Convertible Phaeton			
812 32200	S	FC	2901	C	96	379	Beverly	Sedan (bustleback)	Folding armrests, front and rear	
812 32202	H	FC	2910	C	91	590	Convertible Phaeton			
812 32203	A	FC	2915				Westchester	Sedan (fastback)	Plain	
812 32208	A	FC	2953	C	90	1505	Westchester	Sedan (fastback)	Plain	
812		FC	2961	C	96	498	Beverly	Sedan (bustleback)	Folding armrests, front and rear	
812 32210	H	FC	2919	C	91	683	Convertible Phaeton			
812 32211	S	FC	2918				Beverly	Sedan (bustleback)	Folding armrests, front and rear	
812		FC	2966				Beverly	Sedan (bustleback)	Folding armrests, front and rear	
812 32212	H	FC	2905	C	91	614	Convertible Phaeton			Right hand drive.
812 32213	S	FC	3014				Beverly	Sedan (bustleback)	Folding armrests, front and rear	
812		FC	2940	C	91	540	Convertible Phaeton			
812 2214	F	FB	2243	C	92	284	Convertible Coupe			Exported to Canada.
812		FC	2958				Convertible Phaeton			
812 32215	H	FC	2990	C	91	602	Convertible Phaeton			
812 2216	H	FB	2220	C	91	605	Convertible Phaeton			
812 32217	A	FC	2478				Westchester	Sedan (fastback)	Plain	
812 32218	S	FC	2879				Beverly	Sedan (bustleback)	Folding armrests, front and rear	
812 2219	F	FB	1629				Convertible Coupe			
812		FC	2992	C	96	419	Beverly	Sedan (bustleback)	Folding armrests, front and rear	
812 32220	S	FC	2993				Beverly	Sedan (bustleback)	Folding armrests, front and rear	Exported to Canada.
812 32221	F	FC	2996	C	92	285	Convertible Coupe			
812 32222	S	FC	2994	C	96	415	Beverly	Sedan (bustleback)	Folding armrests, front and rear	
812 32224	H	FC	2998	C	91	607	Convertible Phaeton			
812 32226	F	FC	3007				Convertible Coupe			
812 32227	H	FC	2682	C	91	611	Convertible Phaeton			
812 32229	H	FC	2997	C	91	612	Convertible Phaeton			
812 2231	S	FB	2240				Beverly	Sedan (bustleback)	Folding armrests, front and rear	
812 2232	S	FB	2237	C	96	399	Beverly	Sedan (bustleback)	Folding armrests, front and rear	
812 32233	H	FC	2999	C	91	613	Convertible Phaeton			
812 32234	H	FC	2519				Convertible Phaeton			
812 32236	F	FC	3011				Convertible Coupe			
812 32237	H	FC	3002	C	91	619	Convertible Phaeton			
812 2239	H	FB	2256	C	91	618	Convertible Phaeton			
812 32239	S	FC	3018	C	96	385	Beverly	Sedan (bustleback)	Folding armrests, front and rear	
812 2240	H	FB	1802				Convertible Phaeton			
812 32242	H	FC	3004	C	91	622	Convertible Phaeton			
812 32245	F	FC	2978	C	92	287	Convertible Coupe			
812 32246	H	FC	3020				Convertible Phaeton			Right hand drive, exported to India.
812 2251	S	FB	2202				Beverly	Sedan (bustleback)	Folding armrests, front and rear	
812 32253	H	FC	2675				Convertible Phaeton			
812 32256	H	FC	3017	C	91	643	Convertible Phaeton			
812 2257	S	FB	2228				Beverly	Sedan (bustleback)	Folding armrests, front and rear	
812 2259	S	FB	2258	C	96	421	Beverly	Sedan (bustleback)	Folding armrests, front and rear	
812 2260	S			C	96	422	Beverly	Sedan (bustleback)	Folding armrests, front and rear	
812 32261	F	FC	2995	C	92	288	Convertible Coupe			
812 2262	H	FB	1683				Convertible Phaeton			
812 2263	S	FB	2189	C	96	368	Beverly	Sedan (bustleback)	Folding armrests, front and rear	
812 2264	S	FB	2185	C	96	425	Beverly	Sedan (bustleback)	Folding armrests, front and rear	
812 32265	F	FC	2574	C	92	291	Convertible Coupe			
812 32269	H	FC	2984	C	91	630	Convertible Phaeton			
812 2270	H	FB	2205	C	91	645	Convertible Phaeton			Right hand drive.
812 2273	S	FB	2207	C	96	429	Beverly	Sedan (bustleback)	Folding armrests, front and rear	Right hand drive.
812 32274	F	FC	2987	C	92	290	Convertible Coupe			

Serial #		Engine #		Body #			Model	Body Type	Interior Pattern	Comments
812 32275	S	FC	2988				Beverly	Sedan (bustleback)	Folding armrests, front and rear	
812 2276	A	FB	2246	C	90	1558	Westchester	Sedan (fastback)	Plain	
812 2277	F	FB	2195				Convertible Coupe			
812		FC	3040				Convertible Coupe			
812 32280	H	FC	2980	C	91	655	Convertible Phaeton			
812 2283	S	FB	2178				Beverly	Sedan (bustleback)	Folding armrests, front and rear	
812 2286	S	FB	2748	C	96	451	Beverly	Sedan (bustleback)	Folding armrests, front and rear	
812 32293	H	FC	3037	C	91	639	Convertible Phaeton			
812 32294	A	FC	3025	C	90	1564	Westchester	Sedan (fastback)	Pleated leather	
812 32295	F	FC	3055	C	92	289	Convertible Coupe			Right hand drive.
812 32296	H	FC	3045	C	91	654	Convertible Phaeton			
812 32297	S	FC	3052				Beverly	Sedan (bustleback)	Folding armrests, front and rear	
812 32301	H	FC	3033	C	91	646	Convertible Phaeton			
812 32302	H	FC	3053	C	91	648	Convertible Phaeton			
812 32305	H	FC	3047				Convertible Phaeton			
812 32307	H	FC	3030	C	91	635	Convertible Phaeton			
812 32309	F	FC	3044	C	92	296	Convertible Coupe			
812 32310	H	FC	3039	C	91	627	Convertible Phaeton			
812 2315	A	FB	2198				Westchester	Sedan (fastback)	Plain	
812 32316	A	FC	3031	C	90	1554	Westchester	Sedan (fastback)	Plain	
812 32317	F	FC	3029				Convertible Coupe			
812 32321	H	FC	3027	C	91	657	Convertible Phaeton			
812 32322	F	FC	3062	C	92	305	Convertible Coupe			
812 32325	H	FC	3067	C	91	634	Convertible Phaeton			
812 32327	A	FC	3057	C	96	464	Westchester	Sedan (bustleback)	Plain	Had a bustle trunk, like most Beverlys.
812 2328	S	FB	2214	C	96	479	Beverly	Sedan (bustleback)	Folding armrests, front and rear	
812 2331	H			C	91	669	Convertible Phaeton			
812 2333	A	FB	2226	C	90	1567	Westchester	Sedan (fastback)	Plain	
812				C	90	1574	Westchester	Sedan (fastback)	Plain	
812 2335	S	FB	2217				Beverly	Sedan (bustleback)	Folding armrests, front and rear	
812 32339	H	FC	3079	C	91	695	Convertible Phaeton			
812 32340	H	FC	3075	C	91	652	Convertible Phaeton			
812 2341	S	FB	2865	C	96	485	Beverly	Sedan (bustleback)	Folding armrests, front and rear	
812 2342	S	FB	2872	C	96	494	Beverly	Sedan (bustleback)	Folding armrests, front and rear	
812 2343	S	FB	2485				Beverly	Sedan (bustleback)	Folding armrests, front and rear	
812 2344	A	FB	2241	C	90	1573	Westchester	Sedan (fastback)	Plain	
812 32348	H	FC	3064	C	91	696	Convertible Phaeton			
812		FC	3065				Beverly	Sedan (bustleback)	Folding armrests, front and rear	
812 32350	S	FC	3066				Beverly	Sedan (bustleback)	Folding armrests, front and rear	
812 32352	S	FC	3061	C	96	480	Beverly	Sedan (bustleback)	Folding armrests, front and rear	
812 32353	A						Westchester	Sedan (fastback)	Plain	
812 32354	H	FC	3082	C	91	660	Convertible Phaeton			
812 32355	S	FC	3081				Beverly	Sedan (bustleback)	Folding armrests, front and rear	
812 32356	H	FC	3074	C	91	667	Convertible Phaeton			
812 32358	H	FC	3105	C	91	659	Convertible Phaeton			
812 32359	S	FC	3123	C	96	470	Beverly	Sedan (bustleback)	Folding armrests, front and rear	
812 32362	H	FC	3095	C	91	640	Convertible Phaeton			
812 32363	F	FC	3096				Convertible Coupe			
812 32364	H	FC	3100				Convertible Phaeton			
812 32366	H	FC	3101				Convertible Phaeton			
812 32368	S	FC	3125	C	96	402	Beverly	Sedan (bustleback)	Folding armrests, front and rear	
812		FC	3108				Convertible Phaeton			
812 32369	S	FC	3111	C	96	438	Beverly	Sedan (bustleback)	Folding armrests, front and rear	
812 32370	H	FC	3112	C	91	700	Convertible Phaeton			
812 32371	H	FC	3093	C	91	664	Convertible Phaeton			
812 32374	H	FC	3097	C	91	673	Convertible Phaeton			
812				C	91	678	Convertible Phaeton			
812 32378	S	FC	3110				Beverly	Sedan (bustleback)	Folding armrests, front and rear	
812 32379	S	FC	3119				Beverly	Sedan (bustleback)	Folding armrests, front and rear	
812 32380	S	FC	3080	C	96	486	Beverly	Sedan (bustleback)	Folding armrests, front and rear	
812 32382	A	FC	3046				Westchester	Sedan (fastback)	Plain	
812 2384	S	FB	1939	C	96	453	Beverly	Sedan (bustleback)	Folding armrests, front and rear	
812 2385	H	FB	1476				Convertible Phaeton			
812 32388	H	FC	3157	C	91	642	Convertible Phaeton			
812 2389	A	FB	2250	C	90	1580	Westchester	Sedan (fastback)	Plain	
812 32391	H	FC	3151	C	91	705	Convertible Phaeton			
812 32392	S	FC	3155	C	96	461	Beverly	Sedan (bustleback)	Folding armrests, front and rear	
812 2395	A	FB	2162				Westchester	Sedan (fastback)	Plain	A unique upholstery pattern, with folding armrests front and rear.
812 32398	H	FC	3150	C	91	710	Convertible Phaeton			
812 32399	H	FC	3152				Convertible Phaeton			
812 32400	H	FC	3146	C	91	690	Convertible Phaeton			
812 32401	H	FC	2674				Convertible Phaeton			
812 32402	H	FC	3141				Convertible Phaeton			
812 32404	A	FC	3139				Westchester	Sedan (fastback)	Plain	
812 32405	F	FC	3144	C	92	303	Convertible Coupe			
812		FC	2508				Convertible Coupe			
812 32406	S	FC	3147	C	96	500	Beverly	Sedan (bustleback)	Folding armrests, front and rear	
812 32407	H	FC	3148				Convertible Phaeton			

Serial #		Engine #		Body #			Model	Body Type	Interior Pattern	Comments
812 32408	H	FC	3133	C	91	653	Convertible Phaeton			
812		FC	2961	C	96	498	Beverly	Sedan (bustleback)	Folding armrests, front and rear	
812 32410	S	FC	3137				Beverly	Sedan (bustleback)	Folding armrests, front and rear	
812 32411	S	FC	3130	C	96	483	Beverly	Sedan (bustleback)	Folding armrests, front and rear	
812 32413	S	FC	3188	C	96	501	Beverly	Sedan (bustleback)	Folding armrests, front and rear	
812 32414	S	FC	3189	C	96	492	Beverly	Sedan (bustleback)	Folding armrests, front and rear	
812 32415	S	FC	3182				Beverly	Sedan (bustleback)	Folding armrests, front and rear	
812 32418	H						Convertible Phaeton			
812 32419	H						Convertible Phaeton			
812 32422	H	FC	3160	C	91	691	Convertible Phaeton			
812 32423	H	FC	3161	C	91	708	Convertible Phaeton			
812 32424	H	FC	3158	C	91	663	Convertible Phaeton			
812 2425	S						Beverly	Sedan (bustleback)	Folding armrests, front and rear	
812 32427	M	FC	3165				Hardtop Coupe			Had chrome Auburn headlamps, padded leather top.
812		FB	3219				Convertible Phaeton			
812 2429	S	FB	3220	C	96	456	Beverly	Sedan (bustleback)	Folding armrests, front and rear	
812 2431	H	FB	3222	C	91	672	Convertible Phaeton			
812 2432	A	FB	3221				Westchester	Sedan (fastback)	Plain	
812 2434	H	FB	3215				Convertible Phaeton			
812 32437	A	FC	3183	C	90	1578	Westchester	Sedan (fastback)	Plain	
812 32438	S						Beverly	Sedan (bustleback)	Folding armrests, front and rear	
812 2442	S	FB	3201	C	96	467	Beverly	Sedan (bustleback)	Folding armrests, front and rear	
812 2443	H	FB	3199	C	91	677	Convertible Phaeton			
812 2445	S	FB	3192	C	96	499	Beverly	Sedan (bustleback)	Folding armrests, front and rear	
812 2446	S	FB	3206				Beverly	Sedan (bustleback)	Folding armrests, front and rear	
812 2448	A	FB	3209	C	90	1545	Westchester	Sedan (fastback)	Plain	
812 2450	S	FB	1840				Beverly	Sedan (bustleback)	Folding armrests, front and rear	
812 2453	S	FB	3196	C	96	487	Beverly	Sedan (bustleback)	Folding armrests, front and rear	
812 32454	F	FC	3175	C	92	217	Convertible Coupe			
812 32457	H	FC	3172				Convertible Phaeton			
812		FB	3199				Convertible Phaeton			
812 2459	S	FB	3211	C	96	387	Beverly	Sedan (bustleback)	Folding armrests, front and rear	
812 2460	S	FB	3210	C	96	388	Beverly	Sedan (bustleback)	Folding armrests, front and rear	
812 2461	H	FB	3193	C	91	661	Convertible Phaeton			
812 32462	H	FC	3248	C	91	704	Convertible Phaeton			
812 32463	F	FC	3252	C	92	162	Convertible Coupe			
812 32465	H	FC	3258				Convertible Phaeton			
812 32466	S			C	96	447	Beverly	Sedan (bustleback)	Folding armrests, front and rear	
812 32467	S	FC	3248	C	96	420	Beverly	Sedan (bustleback)	Folding armrests, front and rear	
812 32468	S	FC	3244	C	96	423	Beverly	Sedan (bustleback)	Folding armrests, front and rear	
812 32469	S	FC	3247	C	96	405	Beverly	Sedan (bustleback)	Folding armrests, front and rear	
812 32470	H	FC	3240				Convertible Phaeton			Right hand drive.
812 32471	H	FC	3242	C	91	682	Convertible Phaeton			
812 2473	S	FB	3205	C	96	397	Beverly	Sedan (bustleback)	Folding armrests, front and rear	
812 32474	F	FC	3241	C	92	306	Convertible Coupe			
812 2477	S	FB	2414				Beverly	Sedan (bustleback)	Folding armrests, front and rear	
812 2480	H	FB	1582				Convertible Phaeton			
812 32485	F	FC	3237	C	92	304	Convertible Coupe			Right hand drive, exported to South Africa.
812 32486	A	FC	3245				Westchester	Sedan (fastback)	Plain	Right hand drive.
812 32487	F	FC	3235	C	92	265	Convertible Coupe			Right hand drive.
812 2488	A	FB	3214				Westchester	Sedan (fastback)	Plain	
812 32492	S	FC	3236	C	96	389	Beverly	Sedan (bustleback)	Folding armrests, front and rear	
812 32493	H	FC	3070	C	91	689	Convertible Phaeton			
812 32496	H	FC	3122	C	91	699	Convertible Phaeton			
812 32498	H	FC	3102	C	91	666	Convertible Phaeton			
812 32499	S	FC	3104	C	96	431	Beverly	Sedan (bustleback)	Folding armrests, front and rear	132 wheelbase, but only seven louvers.
812		FC	3231				Convertible Phaeton			
812 2501	S	FB	3297	C	96	414	Beverly	Sedan (bustleback)	Folding armrests, front and rear	
812 2506	S						Beverly	Sedan (bustleback)	Folding armrests, front and rear	
812 2507	S	FB	3279	C	96	391	Beverly	Sedan (bustleback)	Folding armrests, front and rear	
812 32509	H	FC	3232	C	91	697	Convertible Phaeton			
812 32510	F	FC	3253	C	92	308	Convertible Coupe			
812 32515	S	FC	2931	C	96	372	Beverly	Sedan (bustleback)	Folding armrests, front and rear	
812 2516	H	FB	1517	C	91	265	Convertible Phaeton			
812		FC	3310				Beverly	Sedan (bustleback)	Folding armrests, front and rear	
812		FB	3316				Westchester	Sedan (fastback)	Plain	

Model 812, Long Wheelbase

Serial #		Engine #		Body #		Model	Body Type	Interior Pattern	Comments
812 1135	B	FB	1845	C 95	102	Custom Berline	Sedan (fastback)	Berline	Probably a prototype, modified from an 810 sedan. Has a fastback, and short cowl cowl like the 125 wheelbase models.
812 10006	B	FB	1812	C 103	107	Custom Berline	Sedan (bustleback)	Berline	
812 10007	B	FB	2859	C 103	108	Custom Berline	Sedan (bustleback)	Berline	
812 10008	B			C 103	109	Custom Berline	Sedan (bustleback)	Berline	
812 10012	B					Custom Berline	Sedan (bustleback)	Berline	
812 10015	B	FB	2089			Custom Berline	Sedan (bustleback)	Berline	
812 10016	B	FB	1071			Custom Berline	Sedan (bustleback)	Berline	
812 310018	S	FC	2268			Custom Beverly	Sedan (bustleback)	Folding armrests, front and rear	
812 10022	S	FB	2110	C 105	104	Custom Beverly	Sedan (bustleback)	Folding armrests, front and rear	
812 10023	S	FB	2119			Custom Beverly	Sedan (bustleback)	Folding armrests, front and rear	
812 310024	S	FC	2288	C 105	113	Custom Beverly	Sedan (bustleback)	Folding armrests, front and rear	
812 310025	B	FC	2319	C 103	116	Custom Berline	Sedan (bustleback)	Berline	Deluxe rear vanity.
812 10026	S	FB	2117	C 105	106	Custom Beverly	Sedan (bustleback)	Folding armrests, front and rear	
812 310027	S	FC	2284			Custom Beverly	Sedan (bustleback)	Folding armrests, front and rear	
812 310028	S	FB	2147	C 105	112	Custom Beverly	Sedan (bustleback)	Folding armrests, front and rear	
812 310029	B	FC	2293			Custom Berline	Sedan (bustleback)	Berline	Deluxe rear vanity.
812				C 103	118	Custom Berline	Sedan (bustleback)	Berline	Deluxe rear vanity.
812 310032	S	FC	2301	C 101	102	Custom Beverly	Sedan (bustleback)	Folding armrests, front and rear	Windows in the rear quarter panels. Possibly a prototype.
812 310034	S					Custom Beverly	Sedan (bustleback)	Folding armrests, front and rear	
812 310035	B	FC	2275	C 103	125	Custom Berline	Sedan (bustleback)	Berline	Deluxe rear vanity.
812 310040	B			C 103	121	Custom Berline	Sedan (bustleback)	Berline	Deluxe rear vanity.
812 10041	S	FC	2170	C 105	116	Custom Beverly	Sedan (bustleback)	Folding armrests, front and rear	Exported to Canada.
812 10042	S			C 105	118	Custom Beverly	Sedan (bustleback)	Folding armrests, front and rear	
812 10044	B	FB	2345	C 103	122	Custom Berline	Sedan (bustleback)	Berline	
812 310045	S	FC	2310			Custom Beverly	Sedan (bustleback)	Folding armrests, front and rear	
812 310046	S	FC	2311	C 105	122	Custom Beverly	Sedan (bustleback)	Folding armrests, front and rear	
812 10048	S	FB	2418	C 105	124	Custom Beverly	Sedan (bustleback)	Folding armrests, front and rear	
812 10050	S	FB	2193	C 101	103	Custom Beverly	Sedan (bustleback)	Folding armrests, front and rear	Windows in the rear quarter panels. Possibly a prototype.
812 10051	S	FB	2351	C 105	126	Custom Beverly	Sedan (bustleback)	Folding armrests, front and rear	
812 10053	S	FB	2348			Custom Beverly	Sedan (bustleback)	Folding armrests, front and rear	
812 10054	S	FB	2341			Custom Beverly	Sedan (fastback)	Folding armrests, front and rear	
812 10055	S	FB	2794	C 105	130	Custom Beverly	Sedan (bustleback)	Folding armrests, front and rear	
812 310055	S	FC	2405	C 105	130	Custom Beverly	Sedan (bustleback)	Folding armrests, front and rear	
812 310058	S	FC	2409			Custom Beverly	Sedan (bustleback)	Folding armrests, front and rear	
812 10059	S	FB	2339	C 105	136	Custom Beverly	Sedan (bustleback)	Folding armrests, front and rear	
812 10060	S	FB	2340	C 105	135	Custom Beverly	Sedan (bustleback)	Folding armrests, front and rear	
812 10062	S	FB	2714	C 105	192	Custom Beverly	Sedan (bustleback)	Folding armrests, front and rear	
812 10063	S	FB	2349	C 105	140	Custom Beverly	Sedan (bustleback)	Folding armrests, front and rear	
812 10064	S	FB	2335	C 105	141	Custom Beverly	Sedan (bustleback)	Folding armrests, front and rear	
812 310065	S	FC	2400	C 105	142	Custom Beverly	Sedan (bustleback)	Folding armrests, front and rear	
812 10067	S	FB	2334	C 105	143	Custom Beverly	Sedan (bustleback)	Folding armrests, front and rear	
812 310068	S	FC	2527	C 105	125	Custom Beverly	Sedan (bustleback)	Folding armrests, front and rear	
812 310069	B	FC	3001	C 103	131	Custom Berline	Sedan (bustleback)	Berline	Deluxe rear vanity.
812 10071	S	FB	2330			Custom Beverly	Sedan (bustleback)	Folding armrests, front and rear	
812 10073	S	FB	2321			Custom Beverly	Sedan (bustleback)	Folding armrests, front and rear	
812 10075	S	FB	2327			Custom Beverly	Sedan (bustleback)	Folding armrests, front and rear	
812		FB	2336			Custom Beverly	Sedan (bustleback)	Folding armrests, front and rear	
812 10076	S	FB	2375	C 105	145	Custom Beverly	Sedan (bustleback)	Folding armrests, front and rear	
812 10077	S	FB	2367			Custom Beverly	Sedan (bustleback)	Folding armrests, front and rear	
812 10078	B	FB	2364	C 103	123	Custom Berline	Sedan (bustleback)	Berline	
812 10079	S	FB	2366			Custom Beverly	Sedan (bustleback)	Folding armrests, front and rear	
812 10082	S	FB	2420	C 105	155	Custom Beverly	Sedan (bustleback)	Folding armrests, front and rear	
812 10084	S	FB	2435	C 105	151	Custom Beverly	Sedan (bustleback)	Folding armrests, front and rear	
812 310085	B	FC	2514	C 103	125	Custom Berline	Sedan (bustleback)	Berline	Deluxe rear vanity.
812				C 105	146	Custom Beverly	Sedan (bustleback)	Folding armrests, front and rear	
812 10090	S	FB	2421	C 105	161	Custom Beverly	Sedan (bustleback)	Folding armrests, front and rear	
812 10091	S	FB	2452			Custom Beverly	Sedan (bustleback)	Folding armrests, front and rear	
812 10092	S	FB	2458			Custom Beverly	Sedan (bustleback)	Folding armrests, front and rear	
812 10093	S	FB	2462			Custom Beverly	Sedan (bustleback)	Folding armrests, front and rear	
812 10095	S	FB	2448	C 105	166	Custom Beverly	Sedan (bustleback)	Folding armrests, front and rear	
812 310096	S	FC	2733			Custom Beverly	Sedan (bustleback)	Folding armrests, front and rear	
812 10097	S					Custom Beverly	Sedan (bustleback)	Folding armrests, front and rear	
812 10099	S	FB	2433			Custom Beverly	Sedan (bustleback)	Folding armrests, front and rear	
812 310100	B	FC	2511	C 103	129	Custom Berline	Sedan (bustleback)	Berline	Deluxe rear vanity.
812 310105	S	FC	2654	C 105	171	Custom Beverly	Sedan (bustleback)	Folding armrests, front and rear	
812 310109	B	FC	2642	C 103	136	Custom Berline	Sedan (bustleback)	Berline	Deluxe rear vanity.
812 10110	S	FB	2587			Custom Beverly	Sedan (bustleback)	Folding armrests, front and rear	
812 310112	S	FC	2677	C 105	184	Custom Beverly	Sedan (bustleback)	Folding armrests, front and rear	
812 310115	S					Custom Beverly	Sedan (bustleback)	Folding armrests, front and rear	
812 310116	S	FC	2666	C 105	181	Custom Beverly	Sedan (bustleback)	Folding armrests, front and rear	
812 310117	S			C 105	175	Custom Beverly	Sedan (bustleback)	Folding armrests, front and rear	
812 310118	S	FC	2667	C 105	174	Custom Beverly	Sedan (bustleback)	Folding armrests, front and rear	
812 310119	S	FC	2670	C 105	172	Custom Beverly	Sedan (bustleback)	Folding armrests, front and rear	
812 310121	S	FC	2625			Custom Beverly	Sedan (bustleback)	Folding armrests, front and rear	
812 310122	S	FC	2678			Custom Beverly	Sedan (bustleback)	Folding armrests, front and rear	
812 310123	S	FC	2676			Custom Beverly	Sedan (bustleback)	Folding armrests, front and rear	
812 310126	S	FC	2703	C 105	194	Custom Beverly	Sedan (bustleback)	Folding armrests, front and rear	
812 310127	S	FC	2712	C 105	193	Custom Beverly	Sedan (bustleback)	Folding armrests, front and rear	

Serial #		Engine #		Body #		Model	Body Type	Interior Pattern	Comments
812 310128	S	FC	2714	C 105	192	Custom Beverly	Sedan (bustleback)	Folding armrests, front and rear	
812 310133	S	FC	2710	C 105	178	Custom Beverly	Sedan (bustleback)	Folding armrests, front and rear	
812 10134	S	FB	2483	C 105	190	Custom Beverly	Sedan (bustleback)	Folding armrests, front and rear	
812 10135	S	FB	2502	C 105	198	Custom Beverly	Sedan (bustleback)	Folding armrests, front and rear	
812 310141	S	FC	2694	C 105	177	Custom Beverly	Sedan (bustleback)	Folding armrests, front and rear	
812 310146	S	FC	2755	C 105	206	Custom Beverly	Sedan (bustleback)	Folding armrests, front and rear	
812 310147	S	FC	2680	C 105	188	Custom Beverly	Sedan (bustleback)	Folding armrests, front and rear	
812 10149	S	FB	2506	C 105	208	Custom Beverly	Sedan (bustleback)	Folding armrests, front and rear	
812 310153	S	FC	2657			Custom Beverly	Sedan (bustleback)	Folding armrests, front and rear	
812				C 105	213	Custom Beverly	Sedan (bustleback)	Folding armrests, front and rear	
812 310154	S	FC	2684	C 105	203	Custom Beverly	Sedan (bustleback)	Folding armrests, front and rear	
812 310155	S	FC	2810	C 105	235	Custom Beverly	Sedan (bustleback)	Folding armrests, front and rear	
812 310157	S	FC	2736			Custom Beverly	Sedan (bustleback)	Folding armrests, front and rear	
812 10158	S	FB	2555			Custom Beverly	Sedan (bustleback)	Folding armrests, front and rear	
812 310159	S	FC	2613	C 105	216	Custom Beverly	Sedan (bustleback)	Folding armrests, front and rear	
812 310161	S	FC	2777			Custom Beverly	Sedan (bustleback)	Folding armrests, front and rear	
812 10163	S	FB	2171	C 105	219	Custom Beverly	Sedan (bustleback)	Folding armrests, front and rear	
812 310164	S	FC	2779	C 105	220	Custom Beverly	Sedan (bustleback)	Folding armrests, front and rear	
812 10166	B	FB	2598	C 103	142	Custom Berline	Sedan (bustleback)	Berline	Right hand drive, exported to South Africa. Deluxe rear vanity.
812 310167	S	FC	2802			Custom Beverly	Sedan (bustleback)	Folding armrests, front and rear	
812 310171	S			C 105	225	Custom Beverly	Sedan (bustleback)	Folding armrests, front and rear	
812 310172	S	FC	2759	C 105	227	Custom Beverly	Sedan (bustleback)	Folding armrests, front and rear	
812 10173	S	FB	2601			Custom Beverly	Sedan (bustleback)	Folding armrests, front and rear	
812 310174	S	FC	2752	C 105	230	Custom Beverly	Sedan (bustleback)	Folding armrests, front and rear	
812 310175	S	FC	2750	C 105	232	Custom Beverly	Sedan (bustleback)	Folding armrests, front and rear	
812 310176	S	FC	2807			Custom Beverly	Sedan (bustleback)	Folding armrests, front and rear	
812 310178	S	FC	2800	C 105	235	Custom Beverly	Sedan (bustleback)	Folding armrests, front and rear	
812 10179	S	FB	2592			Custom Beverly	Sedan (bustleback)	Folding armrests, front and rear	
812 10180	S	FB	2593	C 105	228	Custom Beverly	Sedan (bustleback)	Folding armrests, front and rear	
812 310181	S	FC	2795			Custom Beverly	Sedan (bustleback)	Folding armrests, front and rear	
812 10182	S	FB	2850	C 105	251	Custom Beverly	Sedan (bustleback)	Folding armrests, front and rear	
812 310186	S	FC	2801			Custom Beverly	Sedan (bustleback)	Folding armrests, front and rear	
812 310189	S	FC	2731			Custom Beverly	Sedan (bustleback)	Folding armrests, front and rear	
812 10194	S					Custom Beverly	Sedan (bustleback)	Folding armrests, front and rear	
812 310196	S	FC	2758	C 105	261	Custom Beverly	Sedan (bustleback)	Folding armrests, front and rear	
812 10198	S	FB	2853	C 105	242	Custom Beverly	Sedan (bustleback)	Folding armrests, front and rear	
812 310199	S	FC	2820	C 105	258	Custom Beverly	Sedan (bustleback)	Folding armrests, front and rear	
812 310201	S	FC	2826			Custom Beverly	Sedan (bustleback)	Folding armrests, front and rear	
812 310205	B	FC	2912	C 103	130	Custom Berline	Sedan (bustleback)	Berline	Deluxe rear vanity.
812 310212	S	FC	2827			Custom Beverly	Sedan (bustleback)	Folding armrests, front and rear	
812 10214	S	FB	2861			Custom Beverly	Sedan (bustleback)	Folding armrests, front and rear	
812 310215	S	FC	2944			Custom Beverly	Sedan (bustleback)	Folding armrests, front and rear	
812 10217	B	FB	2153	C 106	101	Custom Berline	Sedan (bustleback)	Berline	This car has a 135 wheelbase.
812 310219	S	FC	2952			Custom Beverly	Sedan (bustleback)	Folding armrests, front and rear	
812 310220	S	FC	2925	C 105	279	Custom Beverly	Sedan (bustleback)	Folding armrests, front and rear	
812 310222	S	FC	2954			Custom Beverly	Sedan (bustleback)	Folding armrests, front and rear	
812 310224	S	FC	2899	C 105	288	Custom Beverly	Sedan (bustleback)	Folding armrests, front and rear	Right hand drive, exported to South Africa.
812 310225	S	FC	2644	C 105	290	Custom Beverly	Sedan (bustleback)	Folding armrests, front and rear	
812 310226	S	FC	3023			Custom Beverly	Sedan (bustleback)	Folding armrests, front and rear	
812 310227	S	FC	3022			Custom Beverly	Sedan (bustleback)	Folding armrests, front and rear	
812 10228	S	FB	2332	C 105	297	Custom Beverly	Sedan (bustleback)	Folding armrests, front and rear	
812 10230	S			C 105	311	Custom Beverly	Sedan (bustleback)	Folding armrests, front and rear	
812 10231	S	FB	2191	C 105	312	Custom Beverly	Sedan (bustleback)	Folding armrests, front and rear	
812 10234	S	FB	2187	C 105	306	Custom Beverly	Sedan (bustleback)	Folding armrests, front and rear	
812 310238	S					Custom Beverly	Sedan (bustleback)	Folding armrests, front and rear	
812 310241	S	FC	3115	C 105	272	Custom Beverly	Sedan (bustleback)	Folding armrests, front and rear	
812 310242	S	FC	3113	C 105	287	Custom Beverly	Sedan (bustleback)	Folding armrests, front and rear	
812 310244	S	FC	3138	C 105	294	Custom Beverly	Sedan (bustleback)	Folding armrests, front and rear	
812 310245	S	FC	3143			Custom Beverly	Sedan (bustleback)	Folding armrests, front and rear	
812 310246	S			C 105	270	Custom Beverly	Sedan (bustleback)	Folding armrests, front and rear	
812 310247	S	FC	3129	C 105	296	Custom Beverly	Sedan (bustleback)	Folding armrests, front and rear	
812 310248	S			C 105	304	Custom Beverly	Sedan (bustleback)	Folding armrests, front and rear	
812 310249	S	FC	3149	C 105	293	Custom Beverly	Sedan (bustleback)	Folding armrests, front and rear	
812 310252	S	FC	3190	C 105	309	Custom Beverly	Sedan (bustleback)	Folding armrests, front and rear	
812 310254	S	FC	3180	C 105	328	Custom Beverly	Sedan (bustleback)	Folding armrests, front and rear	
812 310256	S	FC	3177			Custom Beverly	Sedan (bustleback)	Folding armrests, front and rear	
812 10257	S	FB	1945	C 105	320	Custom Beverly	Sedan (bustleback)	Folding armrests, front and rear	
812 10264	S	FB	3213	C 105	276	Custom Beverly	Sedan (bustleback)	Folding armrests, front and rear	
812 10265	S	FB	3294	C 105	327	Custom Beverly	Sedan (bustleback)	Folding armrests, front and rear	
812 310266	S	FC	3233			Custom Beverly	Sedan (bustleback)	Folding armrests, front and rear	
812 310268	S	FC	3227			Custom Beverly	Sedan (bustleback)	Folding armrests, front and rear	
812 310269	S	FC	3277			Custom Beverly	Sedan (bustleback)	Folding armrests, front and rear	
812 310270	S	FC	3271	C 105	318	Custom Beverly	Sedan (bustleback)	Folding armrests, front and rear	
812 310272	S	FC	3274	C 105	329	Custom Beverly	Sedan (bustleback)	Folding armrests, front and rear	
812 310275	S	FC	3273			Custom Beverly	Sedan (bustleback)	Folding armrests, front and rear	
812 310276	S	FC	3276			Custom Beverly	Sedan (bustleback)	Folding armrests, front and rear	
812 310277	S	FC	3309	C 105	323	Custom Beverly	Sedan (bustleback)	Folding armrests, front and rear	
812 310279	S	FC	3312	C 105	308	Custom Beverly	Sedan (bustleback)	Folding armrests, front and rear	
812 10280	S	FB	3216			Custom Beverly	Sedan (bustleback)	Folding armrests, front and rear	
812 310281	S	FC	3311	C 105	310	Custom Beverly	Sedan (bustleback)	Folding armrests, front and rear	

APPENDIX V

The charts below show the standard paint and upholstery color combinations available for 1936 and 1937. The information was compiled from factory memos, dealer material and salesmen's color sample books.

As explained in Chapter 12, custom paint colors and exterior/interior combinations were also available.

Color and Upholstery Ensembles for 1936
Westchester and Beverly sedans

Combination number:	30	31	32	33	34	41*
Exterior	Palm Beach Tan	Ganges Green	Black	Cadet Grey	Rich Maroon	Thrush Brown (Light)
Fenders	Palm Beach Tan	Ganges Green	Black	Cadet Grey	Rich Maroon	Thrush Brown (Dark)
Wheels	Palm Beach Tan	Ganges Green	Light Gray	Cadet Grey	Rich Maroon	Thrush Brown (Dark)
Instrument Board	Deep Maroon	Medium Green	Light Gray	Dark Blue	Light Tan	Tan
Garnish mouldings	Deep Maroon	Medium Green	Light Gray	Dark Blue	Light Tan	Tan
Hardware knobs	Palm Beach Tan	Ganges Green	Black	Cadet Grey	Rich Maroon	Thrush Brown (Light)
Steering wheel	Palm Beach Tan	Ganges Green	Black	Cadet Grey	Rich Maroon	Thrush Brown (Light)
Steering wheel post	Palm Beach Tan	Ganges Green	Black	Cadet Grey	Rich Maroon	Thrush Brown (Light)
Remote control unit	Palm Beach Tan	Ganges Green	Black	Cadet Grey	Rich Maroon	Thrush Brown (Light)
Windshield knobs	Palm Beach Tan	Ganges Green	Black	Cadet Grey	Rich Maroon	Thrush Brown (Light)
Glove box handles	Palm Beach Tan	Ganges Green	Black	Cadet Grey	Rich Maroon	Thrush Brown (Light)
Door finishing panels	Palm Beach Tan	Ganges Green	Black	Cadet Grey	Rich Maroon	Thrush Brown (Light)
Robe strap	Palm Beach Tan	Ganges Green	Black	Cadet Grey	Rich Maroon	Thrush Brown (Light)
Rugs	Deep Maroon	Medium Green	Black	Dark Blue	Deep Maroon	Maroon
Upholstery	Deep Maroon	Medium Green	Light Gray	Dark Blue	Light Tan	Tan
Laces & bindings	Palm Beach Tan	Ganges Green	Black	Cadet Grey	Rich Maroon	Maroon

* Beginning in June 1936

Phaetons and Convertible Coupes

Combination number:	35	36	37	38	39	40
Exterior	Palm Beach Tan	Ganges Green	Cadet Grey	Rich Maroon	Black	Black
Wheels	Palm Beach Tan	Ganges Green	Cadet Grey	Rich Maroon	Light Gray	Light Gray
Instrument Board	Deep Maroon	Medium Green	Dark Blue	Light Tan	Light Gray	Light Gray
Garnish mouldings	Deep Maroon	Medium Green	Dark Blue	Light Tan	Light Gray	Light Gray
Hardware knobs	Palm Beach Tan	Ganges Green	Cadet Grey	Rich Maroon	Black	Black
Steering wheel	Palm Beach Tan	Ganges Green	Cadet Grey	Rich Maroon	Black	Black
Steering wheel post	Palm Beach Tan	Ganges Green	Cadet Grey	Rich Maroon	Black	Black
Remote control unit	Palm Beach Tan	Ganges Green	Cadet Grey	Rich Maroon	Black	Black
Glove box handles	Palm Beach Tan	Ganges Green	Cadet Grey	Rich Maroon	Black	Black
Rugs	Deep Maroon	Medium Green	Dark Blue	Deep Maroon	Black	Black
Upholstery	Maroon leather	Green leather	Blue leather	Brown leather	Brown leather	Maroon leather
Top Material	Tan Wexford Top Cloth	Tan Wexford Top Cloth	Tan Wexford Top Cloth	Tan Wexford Top Cloth	Tan Wexford Top Cloth	Tan Wexford Top Cloth

Color and Upholstery Combinations for 1937

Westchester, Beverly and Custom Beverly sedans, and Custom Berlines

Combination number:	30	32	33	34	41	42	45	46
Exterior	Palm Beach Tan	Black	Cadet Grey	Rich Maroon	Thrush Brown (Light)	Clay Rust	Geneva Blue	Cool Orchard Green
Fenders	Palm Beach Tan	Black	Cadet Grey	Rich Maroon	Thrush Brown (Dark)	Clay Rust	Geneva Blue	Cool Orchard Green
Wheels	Palm Beach Tan	Light Gray	Light Gray	Rich Maroon	Thrush Brown (Dark)	Clay Rust	Geneva Blue	Cool Orchard Green
Instrument Board	Deep Maroon	Light Gray	Dark Blue	Light Tan	Tan	Tan	Dark Blue	Medium Green
Garnish mouldings	Deep Maroon	Light Gray	Dark Blue	Light Tan	Tan	Tan	Dark Blue	Medium Green
Hardware knobs	Palm Beach Tan	Black	Cadet Grey	Rich Maroon	Thrush Brown (Light)	Clay Rust	Cadet Grey	Cool Orchard Green
Steering wheel	Palm Beach Tan	Black	Cadet Grey	Rich Maroon	Thrush Brown (Light)	Clay Rust	Cadet Grey	Cool Orchard Green
Steering wheel post	Palm Beach Tan	Black	Cadet Grey	Rich Maroon	Thrush Brown (Light)	Clay Rust	Cadet Grey	Cool Orchard Green
Remote control unit	Palm Beach Tan	Black	Cadet Grey	Rich Maroon	Thrush Brown (Light)	Clay Rust	Cadet Grey	Cool Orchard Green
Windshield knobs	Palm Beach Tan	Black	Cadet Grey	Rich Maroon	Thrush Brown (Light)	Clay Rust	Cadet Grey	Cool Orchard Green
Door finishing panels	Palm Beach Tan	Black	Cadet Grey	Rich Maroon	Thrush Brown (Light)	Clay Rust	Cadet Grey	Cool Orchard Green
Robe strap	Palm Beach Tan	Black	Cadet Grey	Deep Maroon	Thrush Brown (Light)	Maroon	Cadet Grey	Medium Green
Rugs	Deep Maroon	Black	Dark Blue	Deep Maroon	Maroon	Maroon	Dark Blue	Medium Green
Upholstery	Deep Maroon	Light Gray	Dark Blue	Light Tan	Tan	Tan	Dark Blue	Medium Green
Laces & bindings	Palm Beach Tan	Black	Cadet Grey	Rich Maroon	Maroon	Maroon	Cadet Grey	Medium Green

Phaetons and Convertible Coupes

Combination number:	35	37	38	39	40	43	44 *	47 **
Exterior	Palm Beach Tan	Cadet Grey	Rich Maroon	Black	Black	Cigarette Cream	Ivory	Cool Orchard Green
Wheels	Palm Beach Tan	Cadet Grey	Rich Maroon	Light Gray	Light Gray	Cigarette Cream	Ivory	Cool Orchard Green
Instrument board	Deep Maroon	Dark Blue	Light Tan	Light Gray	Light Gray	Red	Black	Medium Green
Garnish mouldings	Palm Beach Tan	Cadet Grey	Rich Maroon	Black	Black	Cigarette Cream	Ivory	Cool Orchard Green
Hardware knobs	Palm Beach Tan	Cadet Grey	Rich Maroon	Black	Black	Cigarette Cream	Ivory	Cool Orchard Green
Steering wheel	Palm Beach Tan	Cadet Grey	Rich Maroon	Black	Black	Cigarette Cream	Ivory	Cool Orchard Green
Steering wheel post	Palm Beach Tan	Cadet Grey	Rich Maroon	Black	Black	Cigarette Cream	Ivory	Cool Orchard Green
Remote control unit	Palm Beach Tan	Cadet Grey	Rich Maroon	Black	Black	Cigarette Cream	Ivory	Cool Orchard Green
Rugs	Deep Maroon	Dark Blue	Deep Maroon	Black	Black	Black	Black	Medium Green
Upholstery	Maroon leather	Blue leather	Brown leather	Brown leather	Maroon leather	Red leather	Black leather	Green leather
Top Material	Tan Wexford Top Cloth	Tan Wexford Top Cloth	Tan Wexford Top Cloth	Tan Wexford Top Cloth	Tan Wexford Top Cloth	Tan Wexford Top Cloth	Tan Wexford Top Cloth	Tan Wexford Top Cloth

* A black top was available on Ivory convertible cars at extra cost.

** This is a guess. Cool Orchard Green replaced Ganges Green just as production of 1937 models began. No combination number is listed for the convertible cars.

APPENDIX VI

Weights and Measures

Some specifications, especially dimensions and weights of the Custom series, were never printed in factory literature. Missing figures are calculated here from factory prints or from measurements of actual cars. These figures are printed in italics.

Weights are from material provided by Auburn to state licensing authorities, and reprinted in dealers' used car price guides. Some of the data sent out by Auburn were obviously estimated. The supercharger and its attendant outside exhaust system appear to have varied in weight between 75 and 187 pounds. There's no weight increase shown for the 812 Beverly over the 810 Beverly, despite the additional weight of the bustle trunk sheet metal.

List prices are FOB Connersville, and do not include transportation or dealer preparation charges. Delivered price on a 1936 Westchester in Philadelphia, for example, was $2,245.

Some Mechanical Specifications

Engine
Type   V-8
Displacement   288.6 cubic inches
Valve arrangement   In block
Main bearings   3
Oil capacity   8 quarts
Cooling system capacity   28 quarts

	Standard	Supercharged
Compression ratio	6.5-1	6.32-1
Firing Order	1L, 3L, 4L, 2L, 2R, 1R, 3R, 4R	1L, 3L, 3R, 2L, 2R, 1R, 4L, 3R
Horsepower		
Advertised	125 @ 3500rpm	170 @ 4200rpm
Actual	117 @ 3600 rpm	186-195 @ 4200rpm
Torque in lb-ft	223 @ 1800	258 @ 2800-273 @ 3000

Transmission gear ratios	1st	2nd	3rd	4th
	2.11	1.36	0.90	0.64
Overall gear ratios:				
Model 810	9.08	5.85	3.88	2.75
Model 812	9.92	6.39	4.24	3.01

Gas tank capacity   21 gallons

Steering gear ratio   18.2-1
Turns lock-to-lock (1936)   3.25
              (1937)   3.9

Turning circle   41 feet

	Westchester Sedan	Beverly Sedan (810)	Beverly Sedan (812)	Convertible Coupe	Convertible Phaeton	Custom Beverly	Custom Berline
Wheelbase	125	125	125	125	125	132	132
Overall length	195.5	195.5	201.5	195.5	195.5	208.5	208.5
Overall width (From factory print)	71	71	71	71	71	71	71
Height, loaded (From factory print)	60.7	60.7	60.7	58.7	58.7	62.2	62.2
Front tread	56	56	56	56	56	56	56
Rear tread	61	61	61	61	61	61	61
Ground clearance (Not including battery box)	9	9	9	9	9	9	9
Weight							
Model 810	3715	3740		3815	3864		
Model 812 standard engine	3715		3740	3833	3900	4088	4115
Model 812 supercharged	3895		3927	3950	4003	4175	4190
Factory Advertised List Prices							
1936	$1,995	$2,095		$2,145	$2,195		
1937	$2,445		$2,545	$2,595	$2,695	$2,960	$3,060
1937 Supercharged	$2,860		$2,960	$3,010	$3,060	$3,375	$3,575

APPENDIX VII

Cord Speed Records

These American Closed-model stock car speed records were set by a supercharged Beverly sedan on a 10-mile circular course at the Bonneville saltbed in Utah on September 16-17, 1937. The runs were monitored and certified by the Contest Board of the American Automobile Association, and are for a flying start in Class "C" and the Unlimited Class.

Distance	Driver	Time	Average MPH
1 Kilometer	Jenkins	20.777	107.66
5 Kilometers	Jenkins	1 : 43.885	107.66
10 Kilometers	Jenkins	3 : 27.771	107.66
50 Kilometers	Jenkins	17 : 25.720	106.96
100 Kilometers	Jenkins	34 : 57.040	106.67
200 Kilometers	Jenkins	1 : 12 : 9.380	103.34
500 Kilometers	Jenkins	2 : 58 : 30.940	104.62
1000 Kilometers	Jenkins-Updike	5 : 59 : 55.040	103.59
2000 Kilometers	Jenkins-Updike-Oliver	12 : 5 : 6.220	102.83
3000 Kilometers	Jenkins-Updike-Oliver	18 : 16 : 36.330	101.99
4000 Kilometers	Jenkins-Updike-Oliver	24 : 25 : 28.120	101.76
25 Kilometers	Jenkins	8 : 40.74	107.39
25 Miles	Jenkins	13 : 59.80	107.17
75 Kilometers	Jenkins	26 : 11.45	106.76
75 Miles	Jenkins	42 : 14.80	106.52
250 Kilometers	Jenkins	1 : 29 : 36.29	104.02
250 Miles	Jenkins	2 : 24 : 29.87	103.81
300 Kilometers	Jenkins	1 : 46 : 59.19	104.54
300 Miles	Jenkins	2 : 58 : 28.66	104.34
400 Kilometers	Jenkins	2 : 23 : 40.26	103.79
400 Miles	Jenkins	3 : 55 : 58.30	104.35
104.650 Miles	Jenkins	1 : 00 0.00	104.65
313.316 Miles	Jenkins	3 : 00 0.00	104.44
621.458 Miles	Jenkins-Updike	6 : 00 0.00	103.58
1233.728 Miles	Jenkins-Updike-Oliver	12 : 00 0.00	102.81
2441.330 Miles	Jenkins-Updike-Oliver	24 : 00 0.00	101.72
1 Mile	Jenkins	33.437	107.66
5 Miles	Jenkins	2 : 47.187	107.66
10 Miles	Jenkins	5 : 32.300	108.34
50 Miles	Jenkins	28 : 6.510	106.73
100 Miles	Jenkins	56 : 23.870	106.39
200 Miles	Jenkins	1 : 54 : 37.620	104.69
500 Miles	Jenkins	4 : 46 : 28.280	104.72
1000 Miles	Jenkins-Updike-Oliver	9 : 42 : 4.020	103.08
2000 Miles	Jenkins-Updike-Oliver	19 : 37 : 17.700	101.93

INDEX

A

Airplane Development Corporation	83
Allen, Ted	44
American Automobile Association	190
Ames, Harold T.	21-23, 73
ATCO	207
Auburn, 1938	183
Auburn Cord Duesenberg (A-C-D) Club	43, 226, 227
Auburn-Central	73, 208
Auburn-Cord News	120
Aviation and Transportation Corporation	205

B

Bartels, Elmer C.	215, 216
Baster, Forrest S.	53, 54, 55
Beal, Hubert	51
Bendix Aviation Company	36
Bendix duo-servo brake	39
Bendix gearshift mechanism	92
Berlines	169, 170, 171, 173
Beverly, 810	48, 140
Beverly, 812	136, 140
Blue Book	223
Body & Art	44, 45, 47
Bonneville Salt Flats	192, 197, 199
Bragg-Kliesrath Division of Bendix Aviation Corporation	66
Budd Company of Philadelphia	43
Buehrig, Gordon Miller	21-28, 41-51, 227
bustle trunk	184
Butler, Don	122

C

cam grind	164, 166
Central Manufacturing Company	105
Citroen	32, 35, 37
Ciuba, Theodore J.	226
clay model	42, 44-49
Columbia Axle of Cleveland	84
Connors, Bill	175
convertible phaeton	185
Cord, Errett Loban	11-19, 73, 201, 202, 209
Cord Car Club	226
Cord Front Drive	91
Cord Owners Club	223
Cordhaven	97, 99, 101
Cosper, Dale	34, 42, 44, 45, 46, 47, 48
Cotter, Bart	44
Crescent Tool and Die Company	105
custom series cars	169

D

Davidson, Lloyd	84, 89
Detroit Gear & Machine Works	34, 64, 69
DeVaux, Norman	207, 210, 211

E

E257	32
E278	32, 33, 38, 40
E294	64-69, 72
E306	32, 66, 68, 69, 70-76, 101
E306 #2	76
Earhart, Amelia	128
Eastman Kodak	70
Eberts, Ken	122, 124
Eclipse Machine Division of Bendix	109
Emanuel, Victor	204

F

F. Joseph Lamb Company	105
Fahlman, Everett G.	53, 57, 58
Faulkner, Roy	72, 73, 103, 112, 115
Firestone Tire and Rubber	193
Fisher Body Corporation	42
Flint Motor Car Company	205
four speed transmission	63, 69, 80, 90
Front drive system	33-40
Front hub	39, 197

G

Gardner, Vince	42-47
Gear Grinding Machine Company	36
Gibbons Tool and Die Company	105
Graham, Joseph	210-213

H

Harris-Leon-Laisne	38
Henie, Sonja	128
Huntington, Roger	198
Hupp Motor Car Company	211

I

Independent Suspension	38
Indianapolis Speedway	190, 193, 197
instrument panel	48
Irwin, Ronald B.	6

J

Jenkins, David Abbott	190, 197, 199, 200
Jenkins, Marvin	197

K

Kublin, George	76-78, 133

L

L-29	33, 34, 38
Landis, Arthur	83, 86
Lavoie, Stanley	33, 37, 44
Limousine Manufacturing Company	60
Long Manufacturing Company	105
Lorenzen, Paul Reuter	42, 44, 45
Ludvigsen, Karl	37, 38
Lycoming V-12	185

M

Madsen, Merril M. 188
Magnavox 49
Manning, L.B. 51, 202, 204
Markin, Morris 202
McInnis, A.H. 115, 116
Menton, Stan 83
Miller, Fred F. 142
Mix, Tom 124, 128, 130
Model 810 ....41, 43, 46, 47, 91

N

Newport, Herb 48, 95, 134, 140, 142

O

Oliver, Bill 195, 197, 198
outside exhaust pipes 167, 168

P

Patterson Motor Car Company 205
Peck, D. Cameron 224
Permold Company, The 57, 58
Pittsburgh Forging Company 88
Post, Dan 169
Powell Pressed Steel Company 88
Practical Die and Stamping Company 105
Pray, Glenn 219, 220, 222
push-button gearshift 92

R

rear drive transmissions 35
Reuben Donnelly Company 120
Rickenbach, August 31, 53-56, 58, 92
Robinson, Richard H. 42, 45
Rotary Company, The 102
Rzeppa, Alfred Hans 36
Rzeppa joint 36, 37, 137, 142

S

Schwitzer-Cummins Corporation 161
See, Lester 204, 205

Sheller Manufacturing Company 70
Slick, Judge 208
Snow, Herbert C. 31-38
Society of Automotive Engineers 190
spare tire sidemounts 185
Spencer Heater Company 52
Spring rates 37
Startix 109
Stein, Waldo 193, 196-198
Stevens Trophy 190, 192, 193, 195, 199, 200
Stromberg EE-15 carburetor 92
Stutz Motor Car Company 207
styling bridge 42

T

Tenite 70
Thomas, Stanley 133
three-speed transmissions 35, 90
Tremulis, Alex 134, 140
trunk space 185
Truscon Steel Company 88

U

United Auto Workers of America 133
Updike, Bert 197, 198

V

V-8 engine 52, 53, 54, 58, 161
valve train 53

W

Warner Gear Works 205
Weaver, Harry 34
Weisheit, Roy E. 71-73, 111
Weiss, Carl W. 36
Weiss, William 102
Weissmuller, Johnnie 128
Willis, P.P. 48
Willys-Overland 208
Winslow, Dallas 201, 205, 206, 207
Wolfer, Otto 192, 193, 196, 198